Continuous Quantum Measurements and Path Integrals

Continuous Quantum Measurements and Path Integrals

Michael B Mensky

P N Lebedev Physics Institute
Russian Academy of Sciences
117924 Moscow, Russia

Institute of Physics Publishing
Bristol and Philadelphia

British Library Cataloguing-in-Publication Data

A catalogue record for this book is available from the British Library.

ISBN 0 7503 0228 3

Library of Congress Cataloging-in-Publication Data are available

941101

Published by IOP Publishing Ltd, a company wholly owned by The Institute of Physics, London.

IOP Publishing Ltd
Techno House, Redcliffe Way, Bristol BS1 6NX, UK

US Editorial Office: IOP Publishing Inc., The Public Ledger Building, Suite 1035, Independence Square, Philadelphia, PA 19106

Typeset by IOP Publishing Ltd in LaTeX
Printed in the UK by Galliard (Printers) Ltd, Great Yarmouth, Norfolk

To my parents

B M and V A Mensky

Table of Contents

Preface

"Quantum measurement" is a professional jargon word meaning "measurement with quantum effects taken into account". This topic has been of great interest, for specialists as well as for the general public, since the beginning of quantum mechanics up the present time. In fact, the interest to this area has grown during the last decade.

From the very beginning the interest in quantum measurements was provoked by their unusual and paradoxical features, such as the impossibility of measuring the position and momentum of an elementary particle simultaneously and with arbitrarily high accuracy. Nowadays this interest continues for both theoretical and practical reasons.

From the practical point of view, even macroscopic measurements (such as the measurement of position of elements of gravitational-wave antenna) have become very precise and require quantum effects to be taken into account. Moreover, in certain conditions these effects are the main restriction on measurement sensitivity.

The theoretical origin of constant and growing interest in quantum measurements is connected with the general interest in the foundations of quantum mechanics, of which the quantum theory of measurements is a central point. Real insight into the principles of quantum mechanics is impossible without an understanding of quantum measurements.

It must be said, though, that from a practical point of view in most cases neither a "deep understanding" of quantum mechanics nor even explicit consideration of quantum measurements is necessary for the quantum mechanical description of real systems and processes. But this is true only for narrow task of the most immediate applications. Anyone who is interested in the first principles of quantum mechanics, or even in more distant applications, inevitably runs into the question of quantum measurements.

The first principles of quantum mechanics are especially important at the present time since quantum theory is undergoing fast development and expansion of its sphere of application. Particularly acute are questions of principle in such new areas of quantum mechanics as quantum gravity and quantum cosmology, the scientific directions developing before our eyes. However, the development of new technology encounters complicated quantum-mechanical problems and also requires deep penetration into the principles of quantum theory.

The subject of the present book is continuous quantum measurements, i.e. measurements prolonged in time. It turned out that the most appropriate tool for describing such a measurement is the Feynman path integral. In the Feynman approach one describes the evolution of a quantum system as though it follows a path, passing through one point after another. This is just like a classical system. However, if for a classical system each individual path forms a complete description of its evolution, for a quantum system only the summation (integration) over all paths according to certain rules describes the evolution.

If a continuous measurement is performed during the evolution of the system, it produces certain information about the path the system takes. It is evident that summation in this case should be made not over all paths but over only those paths compatible with this information. This is the main idea of the path-integral approach to continuous quantum measurements. The technical development of this idea gives formulae for concrete physical effects.

An evident extension of this idea is from continuous measurements of quantum-mechanical systems to continual measurements of quantum fields. The quantum field dynamics can be described by an integral over all field configurations (this integral is often called the path integral too). The dynamics of the same field under a continual (stretched in time and space) measurement should evidently be described by an integral over those field configurations that are compatible with the measurement result (output).

What conclusions can be drawn from the theory of continuous (and continual) measurements? The main one is a specific uncertainty principle for processes (as distinct from the well known uncertainty principle for states). This principle can be formulated in terms of the action $S[q]$ of the physical system and therefore may be called the action uncertainty principle. It states that the history $[q]$ of a system can be traced, with the help of continuous measurement, with an error not less than the error corresponding to the uncertainty of the action δS equal to the quantum of action, $\hbar$.

In practice, this means there is an optimal accuracy for each continuous measurement. If the measurement is rougher than the optimal one, it is inefficient because of the measurement error, as classical measurement theory predicts. If, on the contrary, the measurement is finer than the optimal one, it is inefficient because of large quantum fluctuations (which may be called quantum measurement noise). The optimal regime of measurement is at the boundary between the classical and quantum regimes, and its error places an absolute limit on the precision of a given type of measurement. The only way to overcome this limit is to choose the measurement from the class of so-called quantum nondemolition (QND) measurements, which have no quantum regime and no limit at all.

Many concrete applications of this general statement will be derived, from the sensitivity of the measurement of an oscillator's frequency components to the emergence of time in a quantum Universe as a consequence of its self-measurement. Both practical and theoretical aspects of the theory will be considered.

The book is organized in such a way that its main ideas can be understood without thorough study. Thus some sections or even whole chapters may be skipped. When this is possible without detriment to understanding, the corresponding remarks are made. The minimum possible path through the book is following:

Chapter 1 → Chapter 4 → ... → Chapter 11.

Chapters 5–10 can be selected by the reader according to taste.

The list of references provided in this book is not complete. It contains only those papers that have been used in the author's work. A number of additional references, mostly to books and review papers, have been included when discussing related topics. However, their choice depends on the specific point of view of the author in the subject discussed. Many important books and papers have not been included because their inclusion would require consideration of the same subjects from other points of view, and this was difficult in the framework of the present book. An attempt was made to compensate partly for this incompleteness by including short remarks on the literature. The aim is to give an idea about areas of quantum measurement theory not considered in this book and to provide a preliminary directions to the literature in these areas.

I am indebted to V B Braginsky, who awoke my interest in the theory of precise measurements and specifically in practical aspects of the quantum theory of measurement. V N Rudenko was the first to point out the importance of evaluating quantum effects in continuous measurements. This was a starting point for the investigation, which led later to the path-integral theory of continuous quantum measurements. My deep gratitude is to K S Thorne, who considered my first paper on the subject (see Mensky 1979a) to be interesting and recommended it for *Physical Review*. I am obliged to G A Golubtsova, who was my collaborator and a coauthor of an important paper on quantum-nondemolition measurements. I also had useful discussions of some questions with C M Caves, H Borzeszkowski, R Onofrio, C Presilla, J Halliwell and many other colleagues.

Moscow, Russia
November 13, 1991

Michael B Mensky

1

Introduction to Continuous Quantum Measurements

The main topic of this book is the path-integral theory of continuous quantum measurements. In this introductory chapter we shall expose the principal ideas of this theory on a qualitative level with a minimum of mathematical apparatus.

All physical systems are in fact quantum, but in certain circumstances some of them may approximately be described as classical. This depends on the error with which the action of the system is known (section 1.1). If the system should be considered as a quantum one, then a specific quantum description is necessary and specific quantum features in the behaviour of the system arise. The main distinction in the description of a quantum system is the concept of a probability amplitude (section 1.2), and the principal feature of the quantum system is an uncertainty principle.

A detailed analysis of the concept of an amplitude in the situation when the system undergoes some measurement allows one to obtain a theory of quantum measurement even if the measurement is continuous (prolonged in time). In the latter case, the different paths the system moves along should be considered as alternatives for the motion and characterized by amplitudes (section 1.3).

The uncertainty principle in its well known form $\Delta q \, \Delta p \gtrsim \hbar$ is appropriate to instantaneous measurements. For continuous measurements a modified uncertainty principle can be formulated in terms of the action (section 1.4). According to this principle (in its simplest but weak form) a continuous measurement produces information such that the uncertainty in the action is not less than the quantum of action $\delta S \gtrsim \hbar$.

The reader may skip Chapters 2 and 3, and go directly from this chapter to Chapter 4 without any detriment to understanding of the main points of the theory. Chapter 2 is necessary only for those who have special interest in the link between von Neumann's theory of instantaneous quantum measurements and the path-integral theory of continuous quantum measurements (though the latter can and will be developed quite independently).

Chapter 3 will be useful for a deeper study of the mathematical formalism of path integrals than the level used in Chapter 4.

1.1 QUANTUM AND CLASSICAL SYSTEMS

Quantum mechanics appeared as a theory of microscopic bodies when it had been proved that the motion of microscopic systems cannot be described in the framework of classical physics. However, quantum effects may be important even for macroscopic bodies. The main criterion is in fact inaccuracy in the value of the action S typical for description of the motion in the framework of the given approximation.

The action S is a functional characterizing the dynamics of a system:

$$S[q] = \int_{t'}^{t''} L(q, \dot{q}, t) \, dt.$$

Here L is the Lagrangian of the system, which in the simple case of a one-dimensional mechanical system takes the form

$$L = \frac{1}{2}m\dot{q}^2 - V(t, q),$$

and

$$[q] = \{q(t) \,|\, t' \le t \le t''\}$$

is a path (a trajectory) of the system. It is important that the action functional $S[q]$ may be evaluated not only for the actual path the classical system takes but also for an arbitrary path in the configuration space of the system. In fact, nonclassical paths play a key role in quantum mechanics and specifically in the theory of continuous measurements.

To judge whether the system is quantum or not it is necessary to compare its action with the *Planck constant*, or the *quantum of action*, $\hbar = 1.055 \times 10^{-27}$ erg s.

Let us make this more precise. Any system is in fact a quantum one. However, in an approximate description the quantum features of a certain system may turn out to be negligible. Then this system in this approximation may be considered to be a classical one.

The action of the system provides a quantitative criterion for this. If the errors, characteristic of the given approximation, lead to an indeterminacy ΔS in the action $S[q]$ large compared with the quantum of action, $\Delta S \gg \hbar$, then the system may be considered to be *classical*. If the action $S[q]$ is given with a rather small error, $\Delta S \lesssim \hbar$, then the system needs to be treated as a *quantum* one.

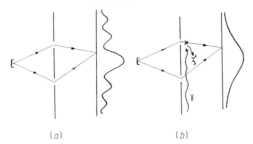

(a) (b)

Figure 1.1: The two-slit experiment leads to an interference pattern if it is not known which slit the particle has passed through (a), but it gives no interference if an additional observation shows which slit was used (b).

1.2 AMPLITUDES AND ALTERNATIVES

From a certain point of view the main object in quantum mechanics is a probability amplitude because it expresses the principal difference between quantum and classical theory.[1] The *probability amplitude* of some event is a complex number A such that $p = |A|^2$ is the probability of this event.

Quantum mechanics differs in that not probabilities but probability amplitudes should be summed up for a quantum system. Suppose that some event can occur through one of two alternative channels and that the probability amplitudes for these channels are A_1 and A_2. Then the complete probability amplitude for the event under consideration is

$$A = A_1 + A_2 \qquad (1.1)$$

and its probability is

$$p = |A|^2 = |A_1 + A_2|^2.$$

A typical example is a particle passing through one of two slits in an opaque screen (figure 1.1), with A_1 being the amplitude for the particle to pass through one of the slits and A_2 that for it to pass through the other. It is because of the law of amplitude summation (1.1) that passing through one of the slits cannot be considered independently from passing through the other slit. The consequence of this law is an interference pattern at the scintillation screen when a series of particles passes through two slits. This phenomenon is discussed in much detail in many textbooks on quantum mechanics (see for example Bohm 1952).

The formula (1.1) is valid, however, only if there is no means of knowing which of two possible alternatives has actually occurred. In this case the

---

[1] Dirac (1972) argued that the most important distinction of quantum theory is not in operators but in amplitudes.

alternatives are called, according to Feynman, *interfering*. They may also be called *quantum alternatives*.

If an additional observation (measurement) is performed giving information about which route has been followed, then the amplitude summation rule changes into the probability summation rule (see Feynman and Hibbs 1965):

$$p = p_1 + p_2 = |A_1|^2 + |A_2|^2.$$

As a consequence, no interference pattern will arise in the two-slit experiment if, for example, the flow of photons falls on the opaque screen so that scattering of the photons shows which of two slits the particle has passed through.[2] In such a situation Feynman called the alternatives *incompatible*. It also seems convenient to use the term *classical alternatives*.

The amplitude summation rule is also valid for many alternative channels,

$$A = A_1 + A_2 + \ldots + A_n \tag{1.2}$$

provided that there is no possibility of discovering which channel has been actually followed. If some observation (measurement) is performed giving information about the channel followed, then the amplitude summation rule must be corrected. The method of correction depends upon the information provided by the measurement.

The information may be complete so that the channel followed is known precisely. Then the probabilities of separate channels should be summed instead of their amplitudes:

$$p = p_1 + p_2 + \ldots + p_n, \qquad p_i = |A_i|^2.$$

For another type of observation information may be only partial. This means that the measurement is rougher. For example, let the measurement permit one to know whether the number of the channel followed belongs to one of the following pairs:

$$(1, 2); \ (3, 4); \ \ldots (n - 1, n)$$

(we suppose that the total number of channels is even). Then the probabilities corresponding to separate pairs are to be summed but amplitudes should be summed inside the pairs:

$$p = p_1 + p_2 + \ldots + p_{n/2}, \qquad p_i = |A_{2i-1} + A_{2i}|^2. \tag{1.3}$$

In this case the alternatives inside each pair are interfering (quantum), while those in different pairs are incompatible (classical) alternatives for

---

[2] Of course, one can say that in this case quite a different physical system is being dealt with. But it is equally valid to talk about an additional observation of the same system. This is typical of the quantum theory of measurement: there is arbitrariness in what part of the real world is included in the system and what is included in the measuring device.

the final event. In the case of yet rougher measurement all channels may be divided into triplets:

$$p = p_1 + p_2 + \ldots + p_{n/3}, \qquad p_i = |A_{3i-2} + A_{3i-1} + A_{3i}|^2. \qquad (1.4)$$

Here p_i are probabilities for different results (outputs) of the measurement. For example, the value

$$p_i = |A_{2i-1} + A_{2i}|^2$$

in equation (1.3) is the probability for the ith pair to emerge as the output of the measurement. The value

$$A_{i\text{th pair}} = A_{2i-1} + A_{2i} \qquad (1.5)$$

is nothing but the probability amplitude for the measurement to give the result expressed by the ith pair. More precisely, this is the amplitude for the event under consideration to occur and the measurement of pairs of channels to show the ith pair.

The situation described by formulae (1.3) and (1.4) can be modelled in a many-slit experiment. The photon flow should then be directed at the opaque screen in such a way that pairs or triplets of slits could be distinguished by the photon scattering rather than individual slits.[3]

1.3 PATHS AND CONTINUOUS MEASUREMENTS

This argument may be applied to *Feynman paths* considered as quantum alternatives. The amplitude $A(q'', q')$ for a particle to move from the point q' to the point q'' is called a *propagator*.[4] It has been expressed by Feynman (1948) in the form of sum (or rather integral) of the amplitudes $A[q]$ corresponding to all possible paths $[q]$ connecting the points q' and q'':

$$A(q'', q') = \int A[q] \, d[q]. \qquad (1.6)$$

Actually this formula is valid for a propagator of any quantum system if q is understood as a coordinate (or a set of coordinates) of the configuration space of this system. For most arguments it is sufficient to consider a one-dimensional system.

---

[3] It might be interesting to perform an experiment of this type for direct experimental investigation of quantum and classical alternatives and the modified amplitude summation rules (1.3) and (1.4).

[4] In the subsequent chapters we shall denote this amplitude by $U(q'', q')$ because it is closely connected with the evolution operator U.

The formula (1.6) is analogous to equation (1.2) but for paths in the role of quantum alternatives. And analogously to the above argument equation (1.6) for the propagator is valid only if there is no possibility of finding out which path is followed when a particle moves from q' to q''. This is usually the case. However, suppose that a *continuous measurement* is performed simultaneously with this transition. Let the output α of this measurement give some information about the path of the transition. Such information can be expressed by some set of paths I_α.[5] If the measurement gives the result (output) α then the transition follows one of the paths $[q]$ belonging to the set I_α. Then, in analogy with equation (1.5), the amplitude for the transition from q' to q'' can be expressed as an integral over paths belonging to I_α:

$$A_\alpha(q'', q') = \int_{I_\alpha} A[q]\, \mathrm{d}[q]. \qquad (1.7)$$

The idea of using restricted path integrals in such a way was proposed in a short remark by Feynman (1948). Some attempts were made to elaborate this idea (see for example Bloch and Burba 1974) but, to our mind, not successfully. The present author, being unaware of this remark of Feynman and subsequent work, proposed this idea again and elaborated it (Mensky 1979a, b, 1983a). In this book other applications of this approach will be considered.[6]

A typical example of continuous measurement is monitoring of the coordinates of the system under consideration. (One may think, for example, of monitoring the position of an elementary particle). Then the measurement gives the value $a(t)$ of the coordinate $q(t)$ at each instant t (of some time interval) with the error Δa determined by the precision of measurement. Then the output of the measurement α can be identified with the path $[a]$ expressed by the curve $a(t)$.

Knowing the output $\alpha = [a]$ of the position monitoring, one knows in fact that the actual path of the system $[q]$ could differ from $[a]$ by no more than the value Δa. Therefore any path $[q]$ lying in the corridor I_α of width $2\Delta a$ around $[a]$ is possible, while no other path is impossible as an actual path of the system (taking the measurement output α into account). Information supplied by the measurement output α is expressed in this case by the corridor I_α of paths. Integration in the Feynman path integral should therefore be performed only over paths in the corridor I_α. Moreover, the corridor I_α may be identified with the output α of the measurement. In fact this corridor represents the output of position monitoring better

---

[5] In fact the set I_α represents the measurement output adequately. This is why we shall later identify the concepts and denote the measurement output and the corresponding set of paths by the same letter α.

[6] The author was probably influenced by Feynman's (1948) paper though unaware of it.

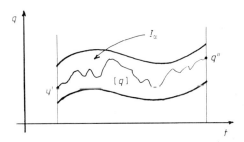

Figure 1.2: Restriction of a path integral to the corridor I_α takes the output α of position monitoring into account.

than the path $[a]$ because it also contains information about the error of measurement (see figure 1.2).[7]

The formula (1.7) will in fact be the basis for all our considerations in this book. If α is fixed, the amplitude $A_\alpha = A_\alpha(q'', q')$ can be considered as the propagator of a particle undergoing continuous measurement (with the given output). If q' and q'' are fixed, the same amplitude can be thought of as a probability amplitude for the continuous measurement to give the result α. Taking a square modulus of the amplitude, one can obtain the probability density for different outputs of the continuous measurement.

A more general class of quantum continuous measurements can be described by the formula

$$A_\alpha(q'', q') = \int w_\alpha[q] A[q] \, d[q]. \tag{1.8}$$

Here integration is performed over all paths connecting the points q' and q'', but with the functional w_α expressing information contained in the output α of the measurement. In the light of this information the system follows a path for which w_α is large. The smaller is $w_\alpha[q]$, the less probable is it that the path $[q]$ is taken by the system.

Generalization of the formulae (1.7) and (1.8) to the case of an arbitrary quantum system and arbitrary continuous measurement is straightforward. Generalization is also possible for the measurement of the configuration of a quantum field. (In the latter case the measurement can be called *continual* because it is not only prolonged in time but also protracted in space.) All these possibilities will be considered in Chapter 4 and the subsequent chapters of this book. However, before more detailed and stricter consideration of the corresponding theory we shall give in section 1.4 a simple formulation of the main results.

---

[7]A track in the Wilson chamber may be thought of as an output of position monitoring. Such a track has a finite width and includes many paths.

1.4 THE ACTION UNCERTAINTY PRINCIPLE

The main and most exciting difference between quantum-mechanical and classical predictions is the Heisenberg uncertainty principle. It can be formulated as follows. If the position q and linear momentum p (of a particle for example) are measured simultaneously, then the indeterminacies Δq and Δp remaining after the measurement should satisfy the inequality

$$\Delta q \, \Delta p \gtrsim \hbar$$

where $\hbar$ is the Planck constant.

The uncertainty principle is good for expressing features of instantaneous quantum measurements but it is inconvenient for continuous measurements. Indeed, if the measurement is, for example, monitoring the coordinate q up to some error Δq, then it gives some information about linear momentum too, but it is difficult to express this information quantitatively and apply the uncertainty principle.

Therefore one must obtain restrictions similar to the uncertainty principle but for continuous measurements. The main objective of this book is the derivation of quantum restrictions on different continuous measurements. In all cases these restrictions will be derived from path integrals of the type of equation (1.7). Let us now analyse some general features of such a calculation and try to give a preliminary formulation of the uncertainty principle for continuous measurements.

The amplitude $A[q]$ of a separate path is, according to Feynman (see for example Feynman and Hibbs 1965),

$$A[q] = e^{\frac{i}{\hbar} S[q]}$$

where $S[q]$ is the action functional for the system under consideration. The path integral (1.7) can therefore be rewritten as

$$A_\alpha = \int_\alpha e^{\frac{i}{\hbar} S[q]} \, d[q] \qquad (1.9)$$

where we have omitted the arguments q' and q'' and identified the output of the measurement α with the corresponding set of paths I_α.

The value A_α can be interpreted as an amplitude for the measurement to give the result (output) α. This means that only those measurement outputs for which the absolute value of A_α is large are probable. The most probable measurement output α_{class} always corresponds to the classical trajectory $[q]_{class}$. However, as a consequence of quantum effects, outputs may differ from α_{class} with a large probability. The uncertainty principle we look for should characterize the whole variance of probable measurement outputs around α_{class}. One can say that it gives a measure of the deviation of the measurement outputs from those predicted by classical theory.

The form of the amplitude (1.9) gives a hint that the corresponding uncertainty principle can be expressed in terms of the action functional $S[q]$. This is why we shall call it *the action uncertainty principle* (ΛUP).

The simplest form of AUP is

$$\delta S \gtrsim \hbar. \tag{1.10}$$

This should be understood in the following sense. The most probable measurement output α_{class} corresponds to the classical theory and therefore to the minimum S_{min} of the action S. Nevertheless outputs α with larger values of the action are probable too, up to the values S, differing from S_{min} by a term of the order of $\hbar$. Thus the variance δS of the action functional S should be not less than the quantum of action $\hbar$ if we calculate this variance over the set of measurement outputs α having comparatively large probabilities.

Remark 1 We denote by Δ(something) the variance of some variable within the limits of the set of paths α, i.e. variance within the limits of a definite measurement output. The notation δ(something) is used for variance within the limits of all outputs α arising with comparatively large probabilities. Thus Δ is connected with the error of the measuring device while δ refers to the variance of measurement outputs.

The action uncertainty principle can be given another form connecting it with the equation of motion. It turns out that the measurement outputs in ideal measurements performed in the quantum regime do not satisfy the classical equation of motion

$$\frac{\delta S[q]}{\delta q(t)} = 0.$$

Instead of this the value of $\delta S[q]/\delta q(t)$ is not zero. The typical deviation of this value from zero can be found from the equality

$$\left| \int_{t'}^{t''} \frac{\delta S[q]}{\delta q(t)} \Delta q(t)\, dt \right| \simeq \hbar. \tag{1.11}$$

Here $[q]$ is the middle path of the set of paths α, and $[\Delta q]$ is a typical deviation of $[q]$ which does not drive it from α, so that $[q + \Delta q] \in \alpha$ too. If this equation is satisfied, then the corresponding measurement output α emerges with comparatively high probability.

Since

$$\frac{\delta S[q]}{\delta q(t)} = \frac{\partial L(q, \dot{q})}{\partial q(t)} - \frac{d}{dt} \frac{\partial L(q, \dot{q})}{\partial q(t)} \tag{1.12}$$

is the left-hand side of the classical equation of motion, equation (1.11) can be symbolically rewritten as follows:

$$\delta(\text{Equation})\Delta(\text{Path}) \simeq \hbar.$$

This means that deviation from the classical picture is insignificant for rough measurements (when $\Delta(\text{Path})$ is large) but is significant for fine measurements (with small $\Delta(\text{Path})$). Fine measurements performed in the quantum regime give rise to nonclassical behaviour of the system as seen in the measurement.

If the left-hand-side (1.12) of a classical equation is not equal to zero, then one could say that a fictitious force $\delta F(t)$ has arisen equal to this nonzero value. Equation (1.11) can therefore be rewritten as

$$\left| \int_{t'}^{t''} \delta F(t) \Delta q(t) \, dt \right| \simeq \hbar. \tag{1.13}$$

The qualitative consequences of this principle in a typical situation can be stated as follows:

1. For a rough measurement (when the error of the measurement is large and the sets α wide) only those measurement outputs predicted by classical theory are probable. This is a classical regime of measurement.

2. For a fine measurement (when the error is small and the sets α narrow) even those measurement outputs far from classical predictions are probable. The more precise the measurement, the wider is the range of probable measurement outputs. This is a quantum regime of measurement.

3. The optimal regime lies between the classical and quantum regimes of measurement. Rougher measurements are inefficient because of classical measurement errors. Finer measurements are inefficient because of the quantum measurement noise.

The variance of outputs of a precise measurement may be considered to be quantum measurement noise. This noise is an additional obstacle to precise measurements. For a certain level of precision this noise becomes the main obstacle. In an attempt to overcome quantum restrictions on the sensitivity of precise measurements Braginsky et al (1977) worked out the concept of so-called quantum nondemolition (QND) measurements (see also Braginsky 1977, Unruh 1979, Caves et al 1980). For a QND measurement there is in fact no quantum regime of measurement. Therefore no quantum threshold arises in this case and there is no absolute restriction on observability. The QND measurement will also be considered in the framework of the path-integral approach and specifically with the help of the action uncertainty principle.

These features of continuous measurements are derived with the help of the action uncertainty principle (AUP) in Chapter 9. However, many other approaches illustrating these aspects of continuous quantum measurements are considered in different chapters. All chapters after Chapter 4 have been made as independent from each other as possible.

2

Instantaneous and Sequential Measurements

Before detailed consideration of continuous measurements we shall first review the main points in the theory of instantaneous measurements.

After a short exposition, in section 2.1, of standard theory that can be found in many textbooks we turn in section 2.2 to an interesting application of this theory to the so-called quantum Zeno paradox (or effect). Finally, in section 2.3, we expand the formalism to the case of approximate measurements of observables with continuous spectra. On the basis of these results, we will consider series of approximate instantaneous measurements, and continuous (prolonged in time) measurements as their limits.

This provides a transition to the main subject of the book—continuous quantum measurements. However, a description of continuous measurements as limits of series of instantaneous measurements turns out to be very cumbersome and in fact inadequate. In subsequent chapters continuous measurements will be introduced in a much simpler way with the help of Feynman path integrals. We will use path integrals as the main tool for investigating continuous measurements. This is why the present chapter may be skipped by the reader who has no special interest in instantaneous measurements and their connection with continuous ones.

2.1 MEASUREMENT OF A QUANTUM SYSTEM

A characteristic feature of quantum systems is that measurements in them unavoidably affect their dynamics. The well known manifestation of this back reaction is the uncertainty principle, discussion of which can be found in any textbook on quantum mechanics. It states that the coordinate q and the linear momentum p of a particle or any other quantum system can be simultaneously measured with errors Δq and Δp restricted by the inequality

$$\Delta q \, \Delta p \gtrsim \hbar. \tag{2.1}$$

In a more direct way the influence of a measurement is revealed in a change of the measured system state. We shall now consider the formalism describing this reaction of a measurement on a quantum system.

2.1.1 Von Neumann's Theory

Let the system be in the (normalized) state $|\psi\rangle$ before measurement, $|\psi\rangle$ denoting a wavefunction $\psi(q)$ or, more generally, a vector of a Hilbert[1] state space $\mathcal{H}$. Let the measurement of an observable A be performed with possible results (outputs) a_i. The observable A is an Hermitian operator in $\mathcal{H}$ and a_i are eigenvalues of this operator, with the eigenstates[2] $|a_i\rangle$:

$$A|a_i\rangle = a_i|a_i\rangle. \tag{2.2}$$

Then, according to von Neumann's postulate (von Neumann 1932), the measurement gives the result a_i with the probability $p_i = |\langle a_i|\psi\rangle|^2$, and the system is in the state $|a_i\rangle$ after the measurement is performed. The number

$$\langle a_i|\psi\rangle = \int \overline{\psi_i(q)}\psi(q)\,\mathrm{d}q \tag{2.3}$$

(where $\psi_i(q)$ is the wavefunction corresponding to the state $|a_i\rangle$) is called a *probability amplitude* for the measurement to give the result a_i, the probability being a square modulus of the probability amplitude. The transition of the system from the state $|\psi\rangle$ into the state $|a_i\rangle$ is called the reduction or collapse of the state $|\psi\rangle$.

If $|\psi\rangle$ can be expanded in the form

$$|\psi\rangle = \sum_i c_i|a_i\rangle$$

then the probability amplitude for the ith measurement output is[3]

$$\langle a_i|\psi\rangle = c_i.$$

This can be expressed in terms of the projectors

$$P_i = |a_i\rangle\langle a_i|. \tag{2.4}$$

The measurement gives the result a_i with the probability $p_i = \langle\psi|P_i|\psi\rangle$ and the system is in the state $P_i|\psi\rangle$ after the measurement.

---

[1] Hilbert space is infinite-dimensional linear space with a scalar product $\langle\psi|\psi'\rangle$.

[2] It is known that these eigenstates are orthogonal as a consequence of A being Hermitian (we suppose all eigenvalues to be different).

[3] The possibility of the above expansion of $|\psi\rangle$ and the formula for an amplitude follow from the fact that the eigenstates $|a_i\rangle$ form a complete set of orthonormal vectors in the Hilbert space $\mathcal{H}$ (the orthonormal basis in $\mathcal{H}$).

In a more general case the measurement may be characterized by a set of projectors[4] $\{P_i\}$ satisfying conditions of completeness and orthogonality:

$$\sum_i P_i = 1, \qquad P_i P_j = 0 \ \text{ for } \ i \neq j \tag{2.5}$$

but otherwise arbitrary. The measurement result (output) corresponding to the projector P_i will happen with probability $p_i = \langle \psi | P_i | \psi \rangle$, leading to collapse of the system into the state $P_i | \psi \rangle$.

2.1.2 Measurements in a Mixed State

Let the initial (before measurement) state of the system be a mixed one characterized by the density matrix ρ. Any density matrix may be presented in the form

$$\rho = \sum_k s_k |\varphi_k\rangle\langle\varphi_k| \tag{2.6}$$

with some set of orthonormal vectors $|\varphi_k\rangle$ and positive coefficients s_k. If the system is in the state ρ one may say instead that it is in one of the states $|\varphi_k\rangle$, with the probability s_k of being in the state $|\varphi_k\rangle$. The coefficients (probabilities) s_k satisfy the relation

$$\sum_k s_k = 1$$

equivalent to the normalization condition for a density matrix,

$$\operatorname{tr} \rho = 1.$$

Let a system in the state ρ be subject to the measurement characterized by the projector set $\{P_i\}$. Then the ith output will be obtained with the probability $p_i = \operatorname{tr}(P_i \rho)$, and the system will collapse (reduce) after this into the state[5]

$$\rho_i = \frac{P_i \rho P_i}{\operatorname{tr}(P_i \rho)}.$$

These two statements may be unified into a single one: the measurement $\{P_i\}$ converts the state ρ into the state

$$\rho' = \sum_i P_i \rho P_i. \tag{2.7}$$

---

[4] That is, Hermitian operators such that $P_i^2 = P_i$.
[5] The expressions for p_i and ρ_i can easily be derived from equation 2.6 and the pure-state formulation of von Neumann's postulate.

Indeed, this density matrix may be presented in the form

$$\rho' = \sum_i p_i \rho_i. \qquad (2.8)$$

This expansion with

$$\text{tr}\,\rho_i = 1 \quad \text{and} \quad \sum_i p_i = 1$$

can be equivalently interpreted as incomplete knowledge about the state. One may say that the system is in one of the states $\{\rho_i\}$ and the probability of the state ρ_i is p_i.

Description of the measurement with the help of the density matrix (2.7) is convenient when it is not known what output is found as a result of the measurement. In this case the measurement is called *non-selective*. Non-selective measurement can occur not only if the measurement has actually been performed but its result is not known, but also if the formula (2.7) is used for *a priori* calculation. Then the output of the measurement cannot be known in principle, so that all possible outputs are to be taken into account. No physical or any other *real* difference between selective and non-selective measurements exists. The only difference is in the way the measurement is described.

In accordance with the above, the collapse or reduction of the state ρ may be interpreted as its conversion into one of ρ_i or into ρ'. These two interpretations are in fact equivalent.

2.1.3 Decoherentization

At first glance, there is an important difference between the two ways of understanding collapse—as conversion of ρ into ρ_i or ρ'. Indeed, collapse into ρ_i means that a single outcome of the measurement is chosen as corresponding to reality, while ρ' contains all possible outcomes. However, describing the final state of the system by ρ' does not mean that in reality no choice is made. Actually some choice is made as soon as the measurement is over. The presence of all possible outputs ρ_i in the expansion (2.8) should be understood as expressing a lack of knowledge. If the measurement has actually been performed, then one of ρ_i has occurred. In some circumstances we do not know the number i. However, even in this case we know the probabilities of different outputs P_i. The best description of the state in these circumstances is with the help of the density matrix ρ'.

Actually an important feature of the collapse (reduction) of a state in the process of measurement is *decoherentization*, i.e. the violation of phase relations between vectors corresponding to alternative measurement outputs. It is convenient to analyse this in terms of pure states. This analysis is in fact quite general because any mixed state (2.6) may be interpreted as describing incomplete knowledge about an actual pure state of the system.

Let us consider again the measurement determined by the operator A or by the system of projectors $\{P_i\}$ with $P_i = |a_i\rangle\langle a_i|$. Let an initial (before the measurement) state of the system be $|\psi\rangle$. The completeness and orthogonality of the projector system (2.5) in the present simple case take the form

$$\sum_i |a_i\rangle\langle a_i| = 1 \qquad (2.9)$$

(expressing the completeness and orthogonality of eigenstates of a Hermitian operator). Using this relation, one can represent the state $|\psi\rangle$ as a superposition

$$|\psi\rangle = \sum_i |a_i\rangle\langle a_i|\psi\rangle = \sum_i c_i |a_i\rangle \qquad (2.10)$$

where the notation

$$c_i = \langle a_i|\psi\rangle$$

is introduced.

The superposition (2.10) is a coherent one in the sense that relative phases $\arg(c_i/c_{i'})$ of the complex coefficients c_i are well defined. The same state $|\psi\rangle$ can be expressed in the form of a density matrix as

$$\rho = |\psi\rangle\langle\psi| = \sum_{i,j} c_i \bar{c}_j |a_i\rangle\langle a_j|. \qquad (2.11)$$

The coherence of the superposition of the states $|a_i\rangle$ is revealed here by the fact that non-diagonal ($i \neq j$) terms are present in the sum (2.11).

When the measurement is performed, the reduction

$$\rho \to \rho' \qquad (2.12)$$

occurs according to the formula (2.7), resulting in

$$\rho' = \sum_i |c_i|^2 |a_i\rangle\langle a_i|.$$

This density matrix differs radically from ρ in that it contains no non-diagonal terms. Therefore ρ' is an incoherent mixture of the states $|a_i\rangle$, that is relative phases of these states are not defined. The reduction (2.12) leads to violation of phase relations, i.e. to *decoherentization*. The coherent superposition of the states $|a_i\rangle$ changes into their incoherent mixture.

The other description of the same reduction expresses decoherentization in another way. One may say that, as a result of reduction, the state $|\psi\rangle$ changes into one of the states $|a_i\rangle$, the ith alternative occurring with the probability $|c_i|^2$. In this formulation the relative phases of the states $|a_i\rangle$

after the measurement have no meaning because only one of these states actually occurs.

There is a difficult question about the physical nature or mechanism of decoherentization. One can also ask what is measurement and what is a measuring device. What are the objective features of this device that distinguish it or its interaction with the measured system from other types of material system and their interactions?

All these and related problems are difficult and not yet completely solved. Readers interested in attempts to construct models of measuring devices can turn, for example, to the work of Peres (1980b), Machida and Namiki (1980, 1984), Zurek (1981, 1982), Walls et al (1985), Joos and Zeh (1985), Zimányi and Károly (1986), Fukuda (1987), Maki (1988) and Unruh and Zurek (1989). Especially convincing are the works of Zurek giving a physical picture of interaction between the measuring device and the measured system resulting in decoherentization.

Other readers, more interested in theoretical analysis of the concept of quantum measurement and of the role of an observer in measurement, can consult almost any textbook on quantum mechanics, for example those of Landau and Lifshits (1958), Bohm (1952), Dirac (1958), Kaempffer (1965), Piron (1976), Blokhintsev (1982, 1987) or one of the large number of works on quantum measurement, for example von Neumann (1932), Wigner (1968), d'Espagnat (1976, 1983), Everett (1957), Wheeler (1957), DeWitt and Graham (1973), Zeh (1988a), Squires (1988) and Rietdijk (1987).

We shall not go deeper into this problem here. However, it seems useful to analyse one of the most intriguing illustrations of the consequences of a quantum measurement reduction: the so-called quantum Zeno paradox.

2.2 QUANTUM ZENO PARADOX

As is well known, the classical Zeno paradox says that an arrow cannot fly because time consists of separate instants and an arrow is at a definite point at any instant. In other words, the fact that an arrow is at a definite point at each instant inhibits its flight (motion). Misra and Sudarshan (1977) gave the name '*quantum Zeno paradox*' to the following statement: frequent (continuous in the limiting case) measurement of a quantum system inhibits its transition into another state. Of course, this statement was derived from quantum mechanics instead of being declared on philosophic grounds, as with the classical Zeno paradox.

This paradox, or rather effect, attracted attention and was discussed in a number of papers: Chiu et al (1977), Ghirardi et al (1979), Peres (1980a), Joos (1984), Kraus (1981), Home and Whitaker (1986) and Castrigiano and Mutze (1984). Another name accepted for this phenomenon in the literature is the *watchdog effect*.

We shall give here a short sketch of the theory of the quantum Zeno effect and of an experiment confirming its existence.

Interest in the Zeno effect was probably connected with the following formulation of it: continuous observation of decay inhibits decaying. It is doubtful that this formulation is valid, and in any case it cannot be experimentally verified now. But the inhibition, caused by continuous observation, of an induced transition from one quantum level to another is a clear consequence of the theory for a system with a discrete spectrum. It is this version of the Zeno effect that has been verified experimentally. Let us consider it shortly.

2.2.1 *Theory of the Effect*

Let the system be originally (at the instant $t_0 = 0$) in the state $|\psi_0\rangle$ and undergo measurements at the instants $t_1, t_2, \ldots$, where $t_i - t_{i-1} = \Delta t$. Let N measurements be performed during the interval T, so that $\Delta t = T/N$. We shall investigate the influence of these measurements on the evolution of the system. Then, as $N \to \infty$, or $\Delta t \to 0$, we can judge about continuous measurement and its influence on evolution. In this limit the Zeno effect should occur.

The measurement will be described by the set of two projectors $P_0 = |\psi_0\rangle\langle\psi_0|$ and $P_1 = 1 - P_0$. For simplicity one can suggest that the system has only one state $|\psi_1\rangle$ orthogonal to $|\psi_0\rangle$, so that $P_1 = |\psi_1\rangle\langle\psi_1|$. Then each measurement answers the question: which of two states is the system in? There are two possible outputs of such a measurement, numbered 0 and 1. The output 0 means that the system is found to be in the state $|\psi_0\rangle$. According to von Neumann's reduction postulate (see section 2.1.1), in this case the state after the measurement is $|\psi_0\rangle$. The other possible output of the measurement is 1. This means that the system is found to be in the state $|\psi_1\rangle$. And it is actually in this state after the measurement (is reduced into this state).

To find the probabilities of both measurement outputs one should find the state of the system immediately before the measurement. This state occurs in the course of free evolution of the system between two measurements.

Let the system be in the state $|\psi_0\rangle$ at the instant $t_0 = 0$ and develop freely (according to its own laws without any measurement) during the interval $[t_0, t_1] = [0, \Delta t]$. Let this evolution (governed by the Schrödinger equation) be described by the superposition

$$|\psi(t)\rangle = c_0(t)|\psi_0\rangle + c_1(t)|\psi_1\rangle \tag{2.13}$$

where $c_0(0) = 1$ and $c_1(0) = 0$. Then the measurement performed at the instant Δt will give the output 1 with probability $p = |c_1(\Delta t)|^2$ and the output 0 with probability $q = 1 - p = |c_0(\Delta t)|^2$. The number p is often called the probability of spontaneous transition $0 \to 1$ during an interval Δt.

If the initial state is $|\psi_1\rangle$, then its evolution is described by the superposition

$$|\psi'(t)\rangle = d_0(t)|\psi_0\rangle + d_1(t)|\psi_1\rangle$$

with $d_0(0) = 0$ and $d_1(0) = 1$. The measurement results (outputs) 0 and 1 at the moment $t_1 = \Delta t$ then occur with probabilities $|d_0(\Delta t)|^2$ and $|d_1(\Delta t)|^2$ correspondingly. These probabilities are expressed[6] by the same numbers: $|d_0(\Delta t)|^2 = p$, $|d_1(\Delta t)|^2 = q = 1 - p$.

Thus if the state is $|\psi_{i_0}\rangle$ at the instant $t_0 = 0$, then the probability of the measurement output i_1 at the instant $t_1 = \Delta t$ is p for $i_1 \neq i_0$ and $q = 1 - p$ for $i_1 = i_0$. After the measurement is over, the system is (due to reduction) in the state $|\psi_{i_1}\rangle$. Then everything repeats in just the same way with free development during the interval $[t_1, t_2] = [\Delta t, 2\Delta t]$ and measurement at the instant $t_2 = 2\Delta t$. The probability of obtaining the same result at the instant t_2 as at the instant t_1 is $q = 1 - p$, and the probability of obtaining the opposite result is p.

It is evident that everything will repeat again and again. Therefore subsequent measurements will give outputs $i_0, i_1, i_2, \ldots, i_N$ with probability

$$P(i_0, i_1, i_2, \ldots, i_N) = p_1 p_2 \ldots p_N$$

where $p_n = p$ if $i_n \neq i_{n-1}$ and $p_n = q = 1 - p$ if $i_n = i_{n-1}$. This product is equal to

$$p^k q^{N-k}$$

if the state changes k times due to measurements and remains unaltered $(N - k)$ times.

A simple calculation shows that the probability for the final state to be the same as the initial state, $i_N = i_0$ (and respectively $i_N \neq i_0$) is equal to

$$P\left(\begin{array}{c}\text{the same}\\\text{opposite}\end{array}\right) = \sum_{\substack{k\\ \text{even}\\ \text{odd}}} \binom{N}{k} p^k q^{N-k}$$

where summation is performed over all even (respectively odd) numbers between 0 and N. The first of these two probabilities can be called the probability of survival of an original state.

---

[6] This is because orthogonal states $|\psi_0\rangle, |\psi_1\rangle$ evolve, due to unitary evolution, into the states $|\psi\rangle, |\psi'\rangle$, which ought to be also orthogonal.

Taking only the first two terms in the formula for the probability of survival of an original state, one has approximately

$$P_{\text{surv}} = P(\text{the same}) = (1-p)^N \left[1 + \frac{N(N-1)}{2} \left(\frac{p}{1-p} \right)^2 \right].$$

If the second term in the square brackets is much less than the first then all the other (omitted) terms are much less still. In this case only the first term is important:

$$P_{\text{surv}} = (1-p)^N. \tag{2.14}$$

This is the case if p is not only small but satisfies the inequality

$$p \ll \frac{1}{N}. \tag{2.15}$$

Let N measurements be performed during the interval $[t_0, t_N] = [0, T]$ so that $\Delta t = T/N$. If for sufficiently small Δt the probability of spontaneous transition $0 \to 1$ is quadratic in Δt,

$$p = \lambda^2 \Delta t^2 = \left(\frac{\lambda T}{N} \right)^2,$$

then the inequality

$$N \gg (\lambda T)^2$$

is satisfied. Equation (2.15) is valid in this case and the formula (2.14) is applicable to give

$$P_{\text{surv}} = \left(1 - \frac{(\lambda T)^2}{N^2} \right)^N = e^{-\lambda^2 T^2 / N}.$$

In the limit $N \to \infty$ when measurement can be considered to be continuous, this probability tends to unity,

$$\lim_{N \to \infty} P_{\text{surv}} = 1.$$

This means that continuous observation (measurement) inhibits the transition $0 \to 1$, which has finite probability if no measurement is performed. This is just what is called the *quantum Zeno effect*.

The condition under which this result was derived is that the probability of spontaneous transition $0 \to 1$ during a small time interval Δt is quadratic in this interval, $p \sim \Delta t^2$. It can be shown that no inhibition (no quantum Zeno effect) emerges for a linear time dependence of probability, $p \sim \Delta t$.

This is why it is difficult (or impossible) to observe the quantum Zeno effect for the decay of an unstable state. The point is that the typical law of spontaneous decay is the exponential law

$$p(t) = 1 - e^{-\lambda t}$$

giving a linear time dependence for small time, $p \sim \Delta t$.

It is true that the exponential law of decay might be violated for very small times, $\Delta t \lesssim \hbar/E$ where E is the energy of the decaying unstable (metastable) state. Therefore, formally one may conclude that Zeno effect could arise if observations are repeated at intervals less than $\hbar/E$.

However, this is in fact impossible technically since this interval is very small (equal to 10^{-21} s for alpha decay). Moreover, such a measurement is hardly feasible in principle. Indeed, conservation of energy with an accuracy E cannot be provided if the time interval between measurements is less than or of the order of $\hbar/E$. This is a consequence of the Bohr–Heisenberg uncertainty principle[7]

$$\Delta E \Delta t \gtrsim \hbar$$

(see Landau and Lifshits 1958). But if energy is not conserved with an accuracy E, then there is no sense in the question whether the system is in an original unstable state or not, and the measurement of the required type is impossible.

This is why the quantum Zeno paradox has been verified experimentally not for spontaneous decay but for an induced transition between two atomic levels, when the probability of transition without measurement has a quadratic time dependence.

2.2.2 Experimental Verification

Different experiments for verification of the quantum Zeno effect have been proposed. Cook (1988) suggested an experiment with a three-level system (an ion in a magnetic trap), and this has since been performed. The idea of the experiment was as follows.

Two levels E_0 and E_1 of the system (see figure 2.1) play the role of the states $|\psi_0\rangle$ and $|\psi_1\rangle$ of section 2.2.1. Transitions between these levels can be induced by radiation having the resonance frequency $\omega_1 = (E_1 - E_0)/\hbar$. The transitions between levels E_0 and E_2 are used to measure the system state: observation of the scattering of radiation in resonance with $\omega_2 = (E_2 - E_0)/\hbar$ allows one to conclude that the system is in the state E_0.

The probability of spontaneous decay from level 1 to level 0 is assumed to be negligible. The influence of 0–1 resonance radiation leads to a coherent superposition of the levels 0 and 1 of the type of (2.13) with coefficients c_0

---

[7] There are, however, papers with arguments against this principle (see for example Vorontsov 1981). It seems that the energy–time uncertainty relation is valid only if measurement is understood in the narrow sense: as a procedure not only giving information about some observable before the measurement but guaranteeing that this observable actually has the same value during the whole time of measurement.

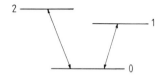

Figure 2.1: A three-level system for verification of the Zeno effect. Repeated observations of whether the system is on level 0 or on level 1 make a transition between them less probable. Level 2 is necessary for the observation.

and c_1 such that

$$p(t) = |c_1(t)|^2 = \frac{1}{2}(1 - \cos \Omega t). \qquad (2.16)$$

Here Ω is the so-called Rabi frequency, proportional to the amplitude of the applied resonance radiation. It is important that for small times Δt the probability is quadratic in time:

$$p(\Delta t) = \frac{1}{4}\Omega^2 \Delta t^2$$

so that the Zeno effect should arise.

Transition from level 2 to level 0 is allowed in the proposed experiment, and level 2 can decay only to level 0. If an intense pulse of 0–2 resonance radiation is applied to the system, this pulse acts as a measurement corresponding to the projectors P_0 and P_1 of section 2.2.1, reducing the system state to one of the states 0 or 1. Indeed, if the system is in the state 0, its transition into state 2 is induced by the action of the resonance radiation (of frequency ω_2), with consequent spontaneous transition back into the state 0. This is accompanied by absorption and consequent emission of the resonance photon, which is equivalent to scattering of the photon. If the system is in the state 1 no scattering occurs. Thus observation of scattering of the 0–2 resonance radiation shows that the system is on level 0, while the absence of scattering means that it is on level 1.

If the system is in a superposition (2.13) of the states 0 and 1, then one of the following events will happen with the corresponding probabilities $|c_1|^2$ and $|c_2|^2$: (1) the resonance photon will be scattered and the system will be found to be in the state 0, or (2) the photon will not be scattered and the system will be in the state 1 after the pulse is over.

Thus the pulse of 0–2 resonance radiation acts as a $\{P_0, P_1\}$ measurement, with reduction to one of the states 0 and 1, and a three-level system of this type is quite appropriate for verification of the Zeno effect. Ideally, an experiment should be protracted as long as $T = \pi/\Omega$. During this time the 0–1 resonance radiation acts on the system (a so-called π-pulse), inducing a coherent superposition of the states 0 and 1, and n short pulses of 0–2 resonance radiation are turned on, separated by equal time intervals T/n. This means that n measurements are performed during the interval T. If $n = 0$, the probability of transition $0 \rightarrow 1$ is given by equation (2.16), and at time T it is equal to unity, so that the system will definitely be in the state 1. However, if $n \neq 0$, the probability of transition $0 \rightarrow 1$ is less than unity, decreasing as n increases. For large n this probability is close to zero, expressing the Zeno effect clearly.

An experiment similar to this has been performed by the group of Wineland (Itano et al 1990). About 5000 ions of beryllium $^9\text{Be}^+$ were stored in a magnetic trap. The levels 0 and 1 were hyperfine sublevels in the ground $2s^2S_{1/2}$ state in a magnetic field. Transition between them was in resonance with the radio frequency 320.7 MHz. Level 2 was one of the sublevels of the $2p^2S_{3/2}$ state, which decays only to level 0. The transition $0 \rightarrow 2$ was in resonance with laser radiation of wavelength 313 nm. The 313 nm fluorescence from the ions was detected, and served as a signal that some ions were in the state 0.

Before the experiment started the ions had been optically pumped[8] to the ground state 0. Then the 0–1 resonance radio-frequency field was applied for $T = 256$ ms, which corresponded to a π-pulse. During this radio-frequency pulse n short (2.4 ms) 313 nm pulses were applied where n was 1, 2, 4, 8, 16, 32 or 64. These pulses were long enough to reduce almost every ion's wavefunction (into one of the states 0 or 1) without causing significant optical pumping.

When the radio-frequency π-pulse was over, the 313 nm laser radiation was turned on to prepare the state 0 again. The number of photons scattered during first 100 ms of this radiation was counted and was approximately proportional to the number of ions at level 0 at the end of the π-pulse (a deviation from proportionality was taken into account with the help of calibration).

The measured numbers of ions at level 0 at the instant T were compared for each n with the numbers predicted by theory (see section 2.2.1). The agreement was almost complete. Thus the theory of the quantum Zeno effect received experimental support. It is interesting that this experimental arrangement reproduces precisely the situation that has been described theoretically by von Neumann's projection (or reduction) postulate.

---

[8] This was done with the help of a long (5 s) 313 nm laser pulse during which those ions that were in level 1 had enough time to decay into level 0 with subsequent repeated induced transitions to level 2 and spontaneous decay into level 0. Finally, when the laser radiation was turned off, almost all ions were at level 0.

2.3 APPROXIMATE AND SEQUENTIAL MEASUREMENTS

The quantum Zeno effect is an example of a continuous measurement giving a trivial result (i.e. resulting in a function that depends on time in a trivial way). However, this is a consequence of the very special character of the measurement procedure described by a discrete set of orthogonal projectors each of which is a bounded operator. In the language of physics, it is important that an observable with a discrete spectrum is considered.

There is of course a wide class of continuous measurements that give non-trivial results. It is just these measurements that will be investigated in the subsequent chapters of this book. A typical example is monitoring of an oscillator coordinate though other systems and measurements will also be considered, among them measurements of quantum field configurations. In all cases the characteristic feature of measurements under consideration will be that the observables to be measured have continuous spectra. Thus our subject will be continuous (prolonged in time) measurements of continuous (having a continuous spectrum) observables.

In Chapter 4 we shall develop the theory of such measurements in the framework of the path-integral formalism. This approach is self-sufficient, so that all necessary formulae will be derived with the help of path integrals. However, it is worthwhile tracing the route from conventional quantum measurement theory sketched above to the path-integral measurement theory. The present section is devoted to this task. It can be skipped without detriment to understanding the rest of the book.

2.3.1 Observables with Discrete versus Continuous Spectra

For simplicity the theory of quantum measurements is usually formulated for observables having discrete spectra. We considered this case above (the eigenvalues a_i in section 2.1.1 were numbered by $i = 1, 2, \ldots$). However, our goal is continuous-spectrum measurements, for example the measurement of position (coordinate) q. One can formulate a simple description of such a measurement in the framework of von Neumann's theory. For this aim one can divide the whole range of the observable q into non-intersecting regions

$$\Delta_i = [q_i, q_{i+1}]$$

and form a new observable Q having discrete values $Q_i = \frac{1}{2}(q_i + q_{i+1})$ when q has a value in Δ_i. Then the measurement of Q can be considered to be a model for the approximate measurement of q, with the error $\Delta q =$

$\frac{1}{2}(q_{i+1} - q_i)$. The observable Q has a discrete spectrum (a discrete set of eigenvalues) Q_i and can be described by a set of orthogonal projectors $\{P_i\}$,

$$(P_i\psi)(q) = \begin{cases} \psi(q) & \text{for } q \in \Delta_i, \\ 0 & \text{otherwise.} \end{cases}$$

It is evident that a real approximate measurement of q differs from this model in two respects. First of all, the result (output) of the approximate measurement can be an arbitrary number q', and this output means that the actual value of q is in the interval

$$\Delta_{q'} = [q' - \Delta q, q' + \Delta q]. \tag{2.17}$$

The crucial difference of this situation from the preceding one is that two intervals $\Delta_{q'}$, $\Delta_{q''}$ intersect if $|q'' - q'| < \Delta q$. If such a measurement is described by a set of projectors $\{P(q')\}$, then the projectors $P(q')$ and $P(q'')$ cannot be orthogonal for $|q''-q'| < \Delta q$. (For example, two projectors $P(q')$ and $P(q'')$ should almost coincide if $|q'' - q'| \ll \Delta q$.) We shall see that a model of this type with non-orthogonal projectors can in fact be constructed. It corresponds to the measurement of q by a measuring device having a rectangular characteristic.

However, the characteristic of a measuring device in most cases is not rectangular but smoother, and this is the cause of the second difference that the measurement of a continuous observable may have as compared with the measurement of a discrete one. The point is that in the above argument all points of an interval Δ_i or $\Delta_{q'}$ were considered on an equal footing. This means that if the measurement gives the result q', then one can conclude that an actual value of q is some point of $\Delta_{q'}$, but no additional information exists.

In real measurements, however, the measurement output q' means that with high probability an actual value of q lies in the middle part of the interval $\Delta_{q'}$. The values close to the ends of this interval are also possible but less probable. Thus there is some distribution (weight function) for the points of the interval $\Delta_{q'}$ like a probabilistic distribution. In fact this distribution may be considered over the whole range of q, so that each measurement output q' can be adequately described by this distribution, or by some measure $\mu_{q'}$, rather than by the interval $\Delta_{q'}$.

This means that the characteristic of the device for measurement of q is not rectangular but a smooth bell-type curve (see figure 2.2), for example the Gaussian curve

$$\Omega_{q'}(q) = J \exp\left(-\frac{(q - q')^2}{2\Delta q^2}\right).$$

We shall see that in the case of a smooth characteristic the measurement cannot be described by a set of projectors. A set of Hermitian positive operators $R(q')$ has to be used instead.

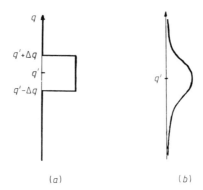

Figure 2.2: The rectangular (a) and smooth (b) characteristics of measuring devices for an observable with a continuous spectrum.

Therefore, to deal with continuous observables instead of discrete ones we should generalize von Neumann's theory of measurement. Let us do this by starting from simultaneous consideration of two discrete-spectrum measurements, one of which is rougher than the other. The rougher measurement will serve as a prototype of the measurement of q with a finite error Δq while the finer measurement will be a prototype of the precise measurement of q.

2.3.2 Approximate Measurements

Consider a pair of discrete-spectrum measurements, one of which is finer than the other. Let $\{P_i\}$ be a set of projectors corresponding to the finer measurement in accordance with von Neumann's scheme (see section 2.1.1). This means that this projector set satisfies the condition of completeness and orthogonality, equation (2.5). Let us call this measurement $\{i\}$-measurement. Consider a rougher measurement ($\{a\}$-measurement) and denote its results (outputs) by the index a. Let the given output a of the rough measurement correspond to the set I_a of the outputs i of the finer $\{i\}$-measurement. This means that performing $\{i\}$-measurement directly after the $\{a\}$-measurement has given the result a allows one to obtain any i from the set I_a but not those i that do not belong to I_a.

The sets I_a and $I_{a'}$ may generally have elements in common, as has been argued in section 2.3.1. Therefore if the $\{a\}$-measurement were defined by a set of projectors $\{P_a\}$, these projectors could not be orthogonal. A tentative choice of these projectors is

$$P_a = \sum_{i \in I_a} P_i.$$

In reality it will be more convenient for us to make use of the operators

$$R_a = \mu P_a \qquad (2.18)$$

proportional to the projectors P_a rather than using the projectors themselves. This is equivalent and differs only by the way of normalization.

Let the system in the state $|\psi\rangle$ undergo $\{a\}$-measurement. Suppose that the measurement output a occurs with the probability p_a and reduces the system into the state $|\psi_a\rangle$.[9] If after this the fine $\{i\}$-measurement is performed, it can give one of the outputs $i \in I_a$ with the probability

$$p_{ai} = \frac{\langle \psi_a | P_i | \psi_a \rangle}{\langle \psi_a | \psi_a \rangle}$$

(this expression is in accordance with von Neumann's theory but for the vector $|\psi_a\rangle$ generally not normalized). The system is reduced in the process of this measurement into the state $P_i|\psi_a\rangle$.

If the system in the same original state $|\psi\rangle$ undergoes the fine $\{i\}$-measurement directly, then the output i emerges, according to von Neumann, with the probability $\langle \psi | P_i | \psi \rangle$ and reduces the system into the state $P_i|\psi\rangle$.

It is natural to suggest that performing the finer measurement after the rougher one leads to the same situation as performing the fine measurement alone. Therefore the description of the two-step measurement

$$|\psi\rangle \xrightarrow{a} |\psi_a\rangle \xrightarrow{i} P_i|\psi_a\rangle \qquad (2.19)$$

is to be in accordance with the description of the direct fine measurement

$$|\psi\rangle \xrightarrow{i} P_i|\psi\rangle. \qquad (2.20)$$

Accordance between corresponding reduced states gives

$$P_i|\psi_a\rangle \sim P_i|\psi\rangle \qquad (2.21)$$

(the norms of the vectors may differ) while agreement between the probabilities requires

$$\sum_{a\,(i \in I_a)} p_a p_{ai} = \langle \psi | P_i | \psi \rangle \qquad (2.22)$$

---

[9]Note that p_a and $|\psi_a\rangle$ do not have to be defined by von Neumann's formulae. Our aim is to derive other formulae applicable in the case of an approximate measurement.

where summation is performed over all values of a such that $i \in I_a$.

The first requirement is satisfied if the reduction under $\{a\}$-measurement has the form

$$|\psi\rangle \xrightarrow{a} |\psi_a\rangle = R_a|\psi\rangle \qquad (2.23)$$

where the operator (2.18) is used. The requirement (2.21) is fulfilled in this case because R_a is proportional to the projector P_a and the latter satisfies the equation

$$P_i P_a = \begin{cases} P_i & \text{for } i \in I_a, \\ 0 & \text{otherwise} \end{cases}.$$

Combining equation (2.22) with equation (2.23) and taking equation (2.5) into account, one has

$$\mu^2 \sum_{a\,(i\in I_a)} \frac{p_a}{\langle\psi|R_a^2|\psi\rangle} = 1.$$

This condition can be satisfied by putting

$$p_a = \langle\psi_a|\psi_a\rangle = \langle\psi|R_a^2|\psi\rangle \qquad (2.24)$$

if $\mu^{-2} = n$ is the number of elements in the set I_a. One can easily see that

$$\sum_a p_a = 1$$

as a consequence of equation (2.5).

Thus we have described a measurement by a set of operators $\{R_a\}$ (proportional to non-orthogonal projectors), with the reduction rule equation (2.23) and probabilities of alternative measurement outputs expressed by equation (2.24). This measurement can be thought of as an approximate measurement of an observable, if the precise its measurement corresponds to the projector set $\{P_i\}$. The error of this approximate measurement is represented by the sets I_a, which are analogues of the intervals $\Delta_{q'}$ discussed in section 2.3.1.

Different points $i \in I_a$ of this 'interval' are considered on an equal footing. This corresponds to a measuring device having a rectangular characteristic. However, this is not realistic. The next step is to introduce a weight function (a non-trivial measurement characteristic) instead of a homogeneous interval.

Let the weight function describing the rough $\{a\}$-measurement be $\mu_{ai} \geq 0$. We shall accept the reduction postulate in the form (2.23) but with

$$R_a = \sum_{i\in I_a} \mu_{ai} P_i. \qquad (2.25)$$

Now the operators R_a are far from being projectors. However, the condition equation (2.21) providing accordance of the reduction operations remains valid as a consequence of

$$P_i R_a = \begin{cases} \mu_{ai} P_i & \text{for } i \in I_a, \\ 0 & \text{otherwise} \end{cases}.$$

The accordance in probabilities (2.22) also takes place if the probability of the measurement output a is determined by equation (2.24) and the following relation is satisfied:

$$\sum_a R_a^2 = 1. \tag{2.26}$$

The latter is provided by

$$\sum_a \mu_{ai}^2 = 1.$$

The operators R_a introduced above are Hermitian positive in the sense that $R_a^\dagger = R_a$ and

$$\langle \psi | R_a | \psi \rangle \geq 0$$

for any state $|\psi\rangle$. We can now forget about the origin of the set $\{R_a\}$ of positive operators. If such a set is given satisfying the normalization condition (2.26), then the measurement is described by the reduction rule (2.23) and the probability distribution (2.24) for different measurement outputs.

In a more general case of a mixed original state ρ the reduction procedure is described as follows:[10]

$$\rho \to \rho_a \;\; = \;\; R_a \rho R_a \tag{2.27}$$

$$p_a \;\; = \;\; \operatorname{tr} \rho_a. \tag{2.28}$$

The generalization of this definition to the case of a continuous-spectrum observable is straightforward. This generalization will be used below to describe (1) a sequential measurement consisting of a series of instantaneous measurements and (2) a continuous (protracted in time) measurement as the limit of such a series.

A more elaborate mathematical formalism for quantum measurements can be found in the works of Davies (1976), Kraus (1983), Holevo (1982), Busch and Lahti (1990) and Busch et al (1990). This formalism describes a

---

[10] Less convenient, in our opinion, though equivalent, is description of the reduction by the normalized density matrices

$$\rho \to \rho_a = \frac{R_a \rho R_a}{\operatorname{tr}(R_a \rho R_a)}, \quad p_a = \operatorname{tr}(R_a \rho R_a).$$

measurement in terms of such mathematical concepts as *instrument, effect* and *operation*. We do not touch this formalism here because the simpler and more 'old-fashioned' description given above seems sufficient and much more transparent. Another way to define the concept we use here is on the basis of the quantum probability theory due to Accardi (1981, 1984).

2.3.3 *Sequential Approximate Measurements*

The description of approximate measurements was derived above for the case when precise (fine) measurement is a discrete-spectrum measurement. However, the resulting formulae can easily be generalized to the case of approximate measurement of continuous-spectrum observables. Take for example the measurement of the position q of a one-dimensional quantum system. In this case the measurement result (output) can be expressed by a real number q'. Let the measurement be performed with the error Δq by a device having a rectangular characteristic so that all points of the interval (2.17) are treated equally. Then the measurement should be described by formulae analogous to equations (2.18), (2.23) and (2.24).

In the present case the system $\{R_{q'}\}$ of operators proportional to projectors should be defined as

$$R_{q'} \;=\; \frac{1}{\sqrt{2\Delta q}} P_{q'}$$

$$(P_{q'}\psi)(q) \;=\; \begin{cases} \psi(q) & \text{for } q \in \Delta_{q'}, \\ 0 & \text{otherwise} \end{cases} .$$

The measurement output q' arises with the probability density

$$p(q') = \langle \psi | R_{q'}^2 | \psi \rangle \tag{2.29}$$

and is followed by the reduction

$$|\psi\rangle \xrightarrow{q'} R_{q'}|\psi\rangle. \tag{2.30}$$

The normalization relation

$$\int_{-\infty}^{\infty} p(q')\,dq' = 1$$

evidently takes place as a consequence of

$$\int_{-\infty}^{\infty} R_{q'}^2\,dq' = 1. \tag{2.31}$$

Consider now a sequence of measurements of this type with intervals of free evolution of the system between the measurements. If the original state at the instant $t_0 = 0$ is $|\psi_0\rangle$, then after the interval $[t_0, t_1]$ (as long as Δt) of free evolution the state is

$$|\psi_1\rangle = U(t_1, t_0)|\psi_0\rangle$$

where $U(t_1, t_0)$ is an evolution operator for the time interval $[t_0, t_1]$. If the measurement of q with the error Δq is performed at the moment t_1, then the output q_1 will be obtained with the probability density

$$p(q_1) = \langle\psi_1|R_{q_1}^2|\psi_1\rangle = \langle\psi_0|U^\dagger(t_1, t_0)R_{q_1}^2 U(t_1, t_0)|\psi_0\rangle.$$

The measurement will result in reduction of the state $|\psi_1\rangle$ as follows:

$$|\psi_1\rangle \xrightarrow{q_1} |\psi_1'\rangle = R_{q_1}|\psi_1\rangle = R_{q_1}U(t_1, t_0)|\psi_0\rangle.$$

The next interval $[t_1, t_2]$ (of duration Δt) results in the evolution of $|\psi_1'\rangle$ into

$$|\psi_2\rangle = U(t_2, t_1)|\psi_1'\rangle.$$

Then the measurement at the instant t_2 gives the result q_2 with the probability density

$$p(q_2) = \frac{\langle\psi_2|R_{q_2}^2|\psi_2\rangle}{\langle\psi_2|\psi_2\rangle}$$

and reduces the state $|\psi_2\rangle$:

$$|\psi_2\rangle \xrightarrow{q_2} |\psi_2'\rangle = R_{q_2}|\psi_2\rangle = R_{q_2}U(t_2, t_1)R_{q_1}U(t_1, t_0)|\psi_0\rangle.$$

Continuing this procedure, we can easily see that after $N - 1$ measurements with outputs $(q_1, q_2, \ldots, q_{N-1})$ and free evolution in the intervals between the measurements we have at the instant $t_N = T$ the state

$$|\psi_N\rangle = U_{q_1 \ldots q_{N-1}}|\psi_0\rangle \tag{2.32}$$

where

$$U_{q_1 \ldots q_{N-1}} = U(t_N, t_{N-1})R_{q_{N-1}} \ldots U(t_2, t_1)R_{q_1}U(t_1, t_0).$$

The probability density of the output $(q_1, q_2, \ldots, q_{N-1})$ of the sequential measurement is

$$p(q_1, q_2, \ldots, q_{N-1}) = p(q_1)p(q_2) \ldots p(q_{N-1}).$$

Taking into account unitarity of the operator $U(t'', t')$ one has for this probability density

$$p(q_1, q_2, \ldots, q_{N-1}) = \langle\psi_N|\psi_N\rangle. \tag{2.33}$$

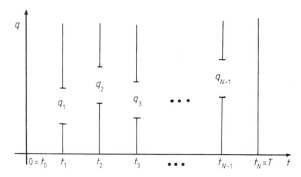

Figure 2.3: The corridor determined by a sequence of approximate measurements of position (coordinate).

The normalization condition

$$\int \cdots \int p(q_1, q_2, \ldots, q_{N-1}) \, dq_1 \, dq_2 \ldots dq_{N-1} = 1$$

can be shown to be satisfied.

In the limit $N \to \infty$, $\Delta t = T/N \to 0$ we have an infinite series of measurements during a finite time interval T. This can be considered to be a model of a *continuous measurement*. The sequence $(q_1, q_2, \ldots, q_{N-1})$ with the interval Δ_{q_i} given for each q_i determines in the limit $N \to \infty$ a continuous corridor (a band) of width $2\Delta q$, see figure 2.3. Let us denote this corridor by α. Then the limit of equation (2.32) is

$$|\psi_\alpha\rangle = U_\alpha|\psi_0\rangle, \quad p_\alpha = \langle\psi_\alpha|\psi_\alpha\rangle \tag{2.34}$$

$$U_\alpha = \lim_{N \to \infty} U_{q_1 \ldots q_{N-1}}. \tag{2.35}$$

The operator (2.35) plays the role of a propagator (probability amplitude) for a quantum system under continuous measurement. It can be expressed in the form of a path integral over paths lying inside the corridor α. This is why path integrals can be used to calculate characteristics of continuous measurements (Mensky 1979a). In the subsequent chapters of the book we shall develop the path-integral formalism for continuous measurements without any reference to instantaneous measurements. However, let us now consider the limit $N \to \infty$ for a more general definition of the instantaneous measurement of q, namely measurement with a smooth characteristic.

An approximate measurement of position q used above is a model corresponding to a rectangular characteristic of a measurement device. If q' is an output of such a measurement, then the actual value of q can be any point of the interval $\Delta_{q'}$ (see (2.17)), each point being equally probable. Real devices correspond to smoother characteristics, such as a Gaussian one. The model of such a measurement should be based upon the set of positive operators $\{R_{q'}\}$, which are not proportional to projectors but rather are analogous to (2.25).

Let us take

$$(R_{q'}\psi)(q) = \Omega(q - q')\psi(q)$$

with $\Omega(x) \geq 0$ being a characteristic of the measuring device. The typical choice is Gaussian:

$$\Omega(x) = J \exp\left(-\frac{x^2}{2\Delta q^2}\right). \tag{2.36}$$

The normalization condition (2.31) is satisfied if

$$\int \Omega^2(x)\,\mathrm{d}x = 1.$$

For the Gaussian characteristic this condition is satisfied if

$$J^2 = \frac{1}{\sqrt{\pi}\Delta q}.$$

With this definition of the operators $R_{q'}$ all formulae derived above for instantaneous and sequential measurements remain valid, including the reduction law (2.30), the probability density for instantaneous measurement outputs (2.29), the evolution under sequential measurement (2.32) and the probability density for the output of the sequential measurement (2.33). However, for transition to the continuous measurement a more complicated limit should be taken now than in equation (2.35). This is because performing several Gaussian measurements consecutively with short time intervals is equivalent to a single Gaussian measurement but of less width Δq (narrower). In order to have a finite limit and a definite width of the effective Gaussian corridor one has to introduce instantaneous measurements with Gaussian characteristics depending on N,[11]

$$\Omega_N(x) = J_N \exp\left(-\frac{x^2}{2\Delta q_N^2}\right)$$

with

$$\Delta q_N^2 = N\Delta q^2.$$

This procedure of taking the limit seems unnatural or at least cumbersome. We shall see in the following chapters that the definition of Gaussian

---

[11] A Gaussian characteristic, though natural and realistic for an instantaneous measurement, does not seem quite appropriate to describe the evolution of a quantum system subject to continuous measurement (see Chapter 10). However, we shall consider this type of measurement because it is mathematically simple.

as well as all other types of continuous measurement becomes quite natural in the framework of the path-integral approach.[12] The corresponding formalism proves to be adequate and efficient. It is important that the two definitions in the simple case considered above lead to equivalent results. However, the definition based on path integrals is more general and applicable even in those cases when presentation of the measurement as a limit of the sequential one is impossible.

The representation of a continuous measurement as a limit of sequential measurement was considered in this chapter because it (1) shows the connection between the concepts necessary for continuous measurements and those for instantaneous measurements, (2) sheds some light on the very concept of continuous measurement, providing an explicit model for it, and (3) makes the main formulae for continuous measurements more convincing for those readers who prefer to have conventional quantum mechanics as a basis for any further development. However, all results derivable from the limit of the sequential measurements (as well as some additional results not derivable in this way) can be obtained much more easily directly from the Feynman path-integral form of quantum mechanics. This will be demonstrated in the rest of the book.

---

[12] The first consideration of continuous quantum measurements was made with the help of the path-integral approach (Mensky 1979a, b). Later on they were defined as limits of sequential measurements by Khalili (1981) and Barchielli et al (1982). The theory of quantum continuous measurements was developed further by Caves (1986, 1987).

3

Technique of Path Integrals

In the late forties Richard Feynman (1948) proposed a path-integral form of quantum mechanics that proved to be equivalent to the conventional Schrödinger-equation form but possessed a much more heuristic force (see Feynman and Hibbs (1965) for a detailed exposition of Feynman's theory). Our consideration of continuous measurements in the following chapters will be based on Feynman path integrals. A theory of quantum continuous measurements will be formulated directly in terms of path integrals instead of a limiting process of the type used in Chapter 2.

The main advantage of the path-integral approach is in its transparent physical sense. However, this approach also has important mathematical advantages, the mathematical aspect of path integrals being elaborated in much detail.

This chapter contains a survey of the path-integral technique as it is used in conventional quantum mechanics (with no measurement taken into account). Readers familiar with this technique can skip this chapter. Those interested only in questions of principle and not in technical detail may read just the first section and skip the rest of the chapter. All readers may skip this chapter in a first reading.

3.1 PROPAGATORS AND PATH INTEGRALS

The state of a quantum system at an instant t can be described by a wavefunction $\psi_t(q)$ depending on the system coordinate (position) q. The latter is a point in some 'configuration space'. This may be the position of a point-like particle, that is a point in the three-dimensional (real) space $\mathbf{R}^3$. However, it may be a point in some multidimensional space if the system under consideration is described by many parameters (coordinates). All the main points of the theory of continuous measurements can be formulated in the simplest case of a one-dimensional system. Therefore q hereafter will be a real number (a point in $\mathbf{R}^1$) unless stated otherwise.

The time dependence of a wavefunction is determined by the *Schrödinger equation*

$$i\hbar\frac{\partial\psi_t}{\partial t} = \hat{H}\psi_t \qquad (3.1)$$

where the quantum Hamiltonian $\hat{H}$ can be obtained from the classical one, $H(p,q,t)$, when $-i\hbar\partial/\partial q$ is substituted for p. Thus knowing the wavefunction ψ_t at the instant t one can solve the Schrödinger equation and find the wavefunction $\psi_{t'}$ at any other instant t'.

The main object in the path-integral approach is a *propagator*, or a probability amplitude for transition from one point q' at an instant t' to another point q'' at an instant t''. We shall denote it by $U(t'',q''|t',q')$. This is a two-point function which may be interpreted as a kernel of the evolution operator $U(t'',t')$. If the propagator is known then the state $|\psi_{t''}\rangle$ can be expressed through $|\psi_{t'}\rangle$ as

$$|\psi_{t''}\rangle = U(t'',t')|\psi_{t'}\rangle$$

or

$$\psi_{t''}(q'') = \int dq' \, U(t'',q''|t',q') \, \psi_{t'}(q'). \qquad (3.2)$$

Remark 1 In many quantum-mechanical problems the Hamiltonian $H = H(p,q)$ does not depend on time explicitly. Then the evolution operator and the propagator depend on t', t'' through the difference $t'' - t'$. In this case

$$U(t'',t') = U(t'' - t') \;=\; e^{-\frac{i}{\hbar}\hat{H}(t''-t')}$$
$$U(t'',q''|t',q') \;=\; \langle q''|e^{-\frac{i}{\hbar}\hat{H}(t''-t')}|q'\rangle$$

with $|q'\rangle$ denoting the state of a definite position, $q = q'$. However, in problems of continuous measurement considered in the subsequent chapters an effective Hamiltonian depends on time explicitly.

The propagator is a solution of the Schrödinger equation

$$i\hbar\frac{\partial}{\partial t''}U(t'',q''|t',q') = \hat{H}''U(t'',q''|t',q') \qquad (3.3)$$

with the delta-function initial condition:

$$U(t',q''|t',q') = \delta(q'' - q'). \qquad (3.4)$$

However, in the path-integral approach the propagator is constructed, independently of the Schrödinger equation, in the form of an integral over all paths connecting the initial and final points:

$$[q] = \{q(t)|\, t' \le t \le t''\} \qquad (3.5)$$

$$q(t') = q', \quad q(t'') = q''. \tag{3.6}$$

Symbolically this can be written as follows:

$$U(t'', q''|t', q') = \int d[q] \exp\left(\frac{i}{\hbar} S[q]\right). \tag{3.7}$$

Here S is an action of the system

$$S[q] = \int_{t'}^{t''} L(q, \dot{q}, t) \, dt$$

expressed through its *Lagrangian* L, which in turn is connected with the Hamiltonian in an ordinary way:

$$H(p, q, t) = p\dot{q} - L(q, \dot{q}, t), \qquad p = \frac{\partial L}{\partial \dot{q}}.$$

The precise definition of a path integral will be given in the next two sections.

An alternative (and often preferable) expression for a propagator is the integral over paths in a phase space. Such a path $[p, q]$ can be defined as a pair consisting of a path $[q]$ in the space of positions (configuration space) and a path

$$[p] = \{p(t)| t' \leq t \leq t''\}$$

in the space of linear momenta. In this case the expression for a propagator is

$$U(t'', q''|t', q') = \int d[p] \int d[q] \exp\left(\frac{i}{\hbar} \int_{t'}^{t''} (p\dot{q} - H(p, q, t)) \, dt\right). \tag{3.8}$$

All the preceding formulas are applicable in quantum mechanics when no measurement is performed during the interval $[t', t'']$ under consideration. We shall see later that in the case of a measurement performed during this interval path integrals should be restricted to appropriate sets of paths. But now let us give a precise definition of path integrals in the formulas (3.7), (3.8).

3.2 DEFINITION OF A PATH INTEGRAL

Path integrals used to construct a quantum system propagator can be defined with the help of *discretization*, or 'skeletonization'. For an integral (3.7) this means that continuous paths $[q]$ should be replaced by continuous piecewise linear curves (lines) of the type drawn in figure 3.1. The

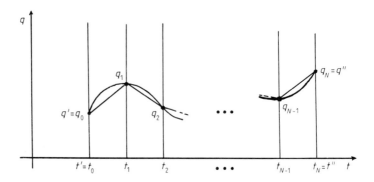

Figure 3.1: Skeletonization of a path $[q]$.

nodes of the line lie on the path $q_i = q(t_i)$. Integration $d[q]$ over continuous paths $[q]$ should be replaced by integration over all possible positions of the nodes,

$$\prod_{i=1}^{N-1} dq_i.$$

This gives an approximation for a path integral. The precise value of the path integral can be obtained in the limit when time interval

$$\Delta t = \frac{(t'' - t')}{N}$$

between nodes of broken lines tends to zero.

The action $S[q]$ of a system should be replaced in the process of skeletonization by the corresponding function of nodes. For the Lagrangian of the form

$$L = \frac{1}{2}m\dot{q}^2 - V(t, q)$$

one has an action

$$S[q] = \int_{t'}^{t''} \left(\frac{1}{2}m\dot{q}^2 - V(t, q) \right) dt \tag{3.9}$$

and skeletonization of an action

$$S(q_0, q_1, \ldots, q_N) = \sum_{i=1}^{N} \left[\frac{1}{2}m \left(\frac{q_i - q_{i-1}}{\Delta t} \right)^2 - V(t_i, q_i) \right] \Delta t. \tag{3.10}$$

Introducing the normalizing factor

$$\prod_{i=1}^{N} \left(\frac{m}{2\pi i \hbar \Delta t} \right)^{1/2}, \tag{3.11}$$

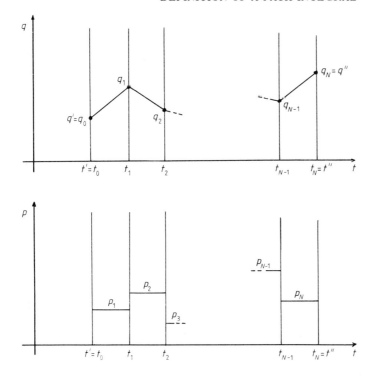

Figure 3.2: Skeletonization of paths $[q]$ and $[p]$.

one has finally

$$U(t'', q''|t', q') = \lim_{N \to \infty} U_{(N)}(t'', q''|t', q')$$

$$= \lim_{N \to \infty} \left(\frac{m}{2\pi i \hbar \Delta t}\right)^{1/2} \int \prod_{i=1}^{N-1} \left(\frac{m}{2\pi i \hbar \Delta t}\right)^{1/2} dq_i$$

$$\times \exp\left[\frac{i}{\hbar} \sum_{i=1}^{N} \left(\frac{1}{2}m \frac{(q_i - q_{i-1})^2}{\Delta t} - V(t_i, q_i)\Delta t\right)\right]. \qquad (3.12)$$

It can be shown that this limit gives a solution to the problem (3.3), (3.4).

The definition (3.12) seems to be not quite satisfactory because of the arbitrary choice of the normalizing factor (3.11). The skeletonization definition of a phase space path integral (3.8) has no such shortcoming. It makes use of the same continuous piecewise linear skeletonization of paths $[q]$ in the configuration space and of the piecewise constant skeletonization of paths $[p]$ in the linear momentum space (figure 3.2).

Then the measure $d[p]d[q]$ should be skeletonized by

$$\prod_{i=1}^{N-1} dq_i \prod_{i=1}^{N} \frac{dp_i}{2\pi\hbar} \qquad (3.13)$$

and the action

$$S[p,q] = \int_{t'}^{t''} \left[p\,dq - \left(\frac{p^2}{2m} + V(t,q) \right) dt \right]$$

should be skeletonized as follows:

$$S(p_1, \ldots, p_N; q_0, q_1, \ldots, q_N)$$
$$= \sum_{i=1}^{N} \left[p_i(q_i - q_{i-1}) - \left(\frac{p_i^2}{2m} + V(t_i, q_i) \right) \Delta t \right].$$

The resulting formula for a propagator is

$$U(t'', q'' | t', q') = \lim_{N \to \infty} U_{(N)}(t'', q'' | t', q') = \lim_{N \to \infty} \int \prod_{i=1}^{N-1} dq_i \prod_{i=1}^{N} \frac{dp_i}{2\pi\hbar}$$
$$\times \exp \left\{ \frac{i}{\hbar} \sum_{i=1}^{N} \left[p_i(q_i - q_{i-1}) - \left(\frac{p_i^2}{2m} + V(t_i, q_i) \right) \Delta t \right] \right\}. \quad (3.14)$$

Using the formula for a so-called *Gaussian integral*,

$$\int_{-\infty}^{\infty} \exp(ax^2 + cx)\,dx = \left(\frac{\pi}{-a} \right)^{1/2} \exp \left(-\frac{c^2}{4a} \right), \qquad (3.15)$$

one can explicitly calculate integrals over all p_i in (3.14). Then the formula (3.12) arises, with the correct normalizing factor. What about the trivial normalizing factor $(2\pi\hbar)^{-1}$ in the integral $dp_i\,dq_i$? It has an evident physical interpretation, since $2\pi\hbar$ is the volume of 'an elementary quantum cell' in the phase space.

Remark 2 We have considered here the case of a one-dimensional system so that p, q are real numbers. For an n-dimensional system when p, q are (real) n-vectors, almost all formulas are valid, but scalar products of vectors should be taken instead of products of numbers, for example

$$p_{(i)}(q_{(i)} - q_{(i-1)}) = \sum_{a=1}^{n} p_{(i)a}(q_{(i)}^a - q_{(i-1)}^a).$$

Besides this, the normalizing factor $(2\pi\hbar)^{-1}$ in (3.14) (correspondingly $(m/2\pi i\hbar\Delta t)^{1/2}$ in (3.12)) should be taken for each degree of freedom, leading to the corresponding measures

$$\prod_{i=1}^{N-1} d^n q_i \prod_{i=1}^{N} \frac{d^n p_i}{(2\pi\hbar)^n} \quad \text{and} \quad \left(\frac{m}{2\pi i\hbar\Delta t} \right)^{n/2} \prod_{i=1}^{N-1} \left(\frac{m}{2\pi i\hbar\Delta t} \right)^{n/2} d^n q_i. \quad (3.16)$$

3.3 THE PATH INTEGRAL FOR AN OSCILLATOR

The most difficult challenge for the theory of path integrals is their evaluation for different potentials $V(t, q)$. There are four ways to solve this:

1. There are methods of exact evaluation of path integrals for certain classes of non-quadratic potentials.

2. For a quadratic potential with a non-quadratic additional term (perturbation) the path integral can be evaluated approximately with the help of perturbation theory.

3. One can use the fact that a Feynman path integral satisfies the Schrödinger equation and construct it as a solution to this equation.

4. The path integral can be evaluated for any quadratic potential and more generally for any quadratic action.

We will not examine the first possibility here, and refer readers to the papers of Duru and Kleinert (1982), Duru (1984), Bin Kang Cheng (1989), Grosche (1989), Anderson and Anderson (1990) and references therein. The second possibility will be outlined briefly in the next section and the third will be considered in Chapter 4. Here let us consider an example of the last approach: an exactly solvable quadratic potential (a more general and in a sense a simpler consideration will be given in the next section).

First of all consider a free particle with the Hamiltonian

$$H = \frac{p^2}{2m}.$$

In this case the skeletonized path integral (3.12) takes the form

$$U_{(N)}(t'', q''|t', q') = \int dq_1 \int dq_2 \ldots \int dq_{N-2} \int d_{N-1}$$
$$U_0(t'', q''|t_{N-1}, q_{N-1})\, U_0(t_{N-1}, q_{N-1}|t_{N-2}, q_{N-2}) \times \ldots$$
$$\times U_0(t_2, q_2|t_1, q_1)\, U_0(t_1, q_1|t', q') \qquad (3.17)$$

with

$$U_0(t'', q''|t', q') = \left(\frac{m}{2\pi i\hbar(t'' - t')}\right)^{1/2} \exp\left(\frac{im}{2\hbar}\frac{(q'' - q')^2}{t'' - t'}\right). \qquad (3.18)$$

It can easily be verified with the aid of the Gaussian integral (3.15) that the kernel U_0 satisfies the so-called *Einstein–Smoluchowski condition* (the *Chapman–Kolmogorov condition* as mathematicians prefer to call it):

$$\int dq' \, U_0(t'', q''|t', q') \, U_0(t', q'|t, q) = U_0(t'', q''|t, q). \qquad (3.19)$$

Therefore all integrals in (3.17) can be evaluated, resulting in

$$U_{(N)}(t'', q''|t', q') = U_0(t'', q''|t', q').$$

Taking the limit $N \to \infty$ and making use of (3.12) one has then

$$U(t'', q''|t', q') = U_0(t'', q''|t', q'),$$

and thus (3.18) is an explicit expression for the free particle propagator.

Now let our system be an oscillator of frequency ω driven by an external force $F(t)$:

$$H = \frac{p^2}{2m} + \frac{m\omega^2 q^2}{2} - F(t)q \qquad (3.20)$$

$$L = \frac{m\dot{q}^2}{2} - \frac{m\omega^2 q^2}{2} + F(t)q \qquad (3.21)$$

$$S[q] = \int_{t'}^{t''} \left(\frac{m\dot{q}^2}{2} - \frac{m\omega^2 q^2}{2} + F(t)q \right) dt. \qquad (3.22)$$

Denote by $q_{\text{class}}(t)$ the classical trajectory of the system, i.e. a solution to the problem

$$m\ddot{q}_{\text{class}} + m\omega^2 q_{\text{class}} = 0$$

$$q_{\text{class}}(t') = q', \quad q_{\text{class}}(t'') = q''.$$

Let us present an arbitrary path $[q]$ between the points q', q'' in the form of a sum

$$q(t) = q_{\text{class}}(t) + z(t) \qquad (3.23)$$

of the classical trajectory and a path $[z]$ having null boundary conditions, $z(t') = z(t'') = 0$. Then integration by parts shows that the action (3.22) is a sum

$$S[q] = S[q_{\text{class}}] + S_0[z]$$

where S_0 is the action of a free oscillator obtained from (3.22) by putting $F \equiv 0$:

$$S_0[z] = \frac{m}{2} \int_{t'}^{t''} (\dot{z}^2 - \omega^2 z^2) \, dt.$$

The classical path $[q_{\text{class}}]$ being unambiguously determined, integration over paths $[q]$ in (3.7) is reduced to integration over paths $[z]$. The formula (3.7) with the action (3.22) then takes the form

$$U(t'', q''|t', q') = J \exp\left(\frac{i}{\hbar} S[q_{\text{class}}]\right) \tag{3.24}$$

where the factor J is defined by the integral

$$J = \int_0^0 d[z] \exp\left(\frac{im}{2\hbar} \int_{t'}^{t''} (\dot{z}^2 - \omega^2 z^2) \, dt\right) \tag{3.25}$$

over paths with null boundary conditions.

The classical action $S[q_{\text{class}}]$ can be evaluated explicitly, so that the formula (3.24) determines the dependence of the propagator U on the points q', q''. The factor J depends only on the parameters m, ω and the length

$$T = t'' - t'$$

of the time interval. This factor can be calculated with the help of a spectral representation (Feynman and Hibbs 1965) of the integral (3.25).

In order to do this let us take the expansion of the path $[z]$ into a Fourier series

$$z(t) = \sum_{n=1}^{\infty} z_n \sin \Omega_n (t - t') \tag{3.26}$$

where we introduce the notation

$$\Omega_n = \frac{n\pi}{T}.$$

Then integration over $[z]$ is reduced to integration over spectral components $z_1, z_2, \ldots$. With the help of the *Parseval formula*

$$\|u\|^2 = \int_{t'}^{t''} |u(t)|^2 \, dt = \frac{T}{2} \sum_{n=1}^{\infty} |u_n|^2 \tag{3.27}$$

which is valid for expansion of any (generally complex) function $u(t)$ into a series of the type of (3.26), one can express the action $S_0[z]$ in terms of the spectral components z_n:

$$S_0[z] = \frac{mT}{4} \sum_{n=1}^{\infty} z_n^2 (\Omega_n^2 - \omega^2).$$

As a result, the integral (3.25) takes the form

$$J = J_0 \int dz_1 \, dz_2 \ldots \exp\left(\frac{imT}{4\hbar} \sum_{n=1}^{\infty} z_n^2 (\Omega_n^2 - \omega^2)\right) \tag{3.28}$$

with J_0 being a numerical factor that does not depend on ω.

The infinite-multiple integral over z_n is in fact a product of one-dimensional integrals of Gaussian type. They can be evaluated with the help of equation (3.15) to give

$$J = J_0 \prod_{n=1}^{\infty} \left(\frac{4\pi i\hbar}{mT(\Omega_n^2 - \omega^2)} \right)^{1/2} .$$

Making use of the *Euler infinite product*,

$$\prod_{n=1}^{\infty} \left(1 - \frac{\omega^2}{\Omega_n^2} \right) = \frac{\sin \omega T}{\omega T}, \tag{3.29}$$

and finding J_0 from the requirement for the propagator (3.24) to coincide with equation (3.18) in the limit $\omega \to 0$, one has finally

$$J = \left(\frac{m\omega}{2\pi i\hbar \sin \omega T} \right)^{1/2} .$$

Using this expression in equation (3.24) one obtains the following formula for the propagator of a driven oscillator:

$$U(t'', q''|t', q') = \left(\frac{m\omega}{2\pi i\hbar \sin \omega T} \right)^{1/2} \exp \left(\frac{i}{\hbar} S[q_{\text{class}}] \right) . \tag{3.30}$$

3.4 GAUSSIAN PATH INTEGRALS

The path integral for an harmonic oscillator evaluated in the preceding section is an example of a Gaussian path integral. We will now briefly consider Gaussian path integrals from a more general point of view. A more detailed discussion of path integrals can be found in Feynman and Hibbs (1965) and Itzykson and Zuber (1980). [1]

The starting point for this consideration is a one-dimensional *Gaussian integral*

$$\int_{-\infty}^{\infty} dq \exp \left(-\frac{1}{2}q^2 + cq \right) = (2\pi)^{1/2} \exp \left(\frac{1}{2}c^2 \right) . \tag{3.31}$$

---

[1] A mathematically strict theory of path integrals can be found in Kac (1958), but it deals mostly with the Wiener path integrals arising in classical statistical mechanics (these integrals are characterized by Hermitian operators A in the notation to be introduced below in the present section). There are many papers devoted to a mathematically strict foundation of Feynman path integrals arising in quantum mechanics (anti-Hermitian operators A), see for example Morette-DeWitt (1972), Klauder (1987) and references therein. (It is argued, however, in Mensky (1992d) that replacement of the concept of an isolated system by a physically more realistic picture of the system measured by its environment leads to the well defined restricted Feynman integrals.)

Taking a product of n integrals of this type, one has in fact a Gaussian integral over n-dimensional vectors:

$$\int \prod_{i=1}^{n} dq_i \, \exp\left(-\frac{1}{2}(q, q) + (c, q)\right) = (2\pi)^{n/2} \exp\left(\frac{1}{2}(c, c)\right)$$

where a scalar product is introduced:

$$(c, q) = \sum_{i=1}^{n} c_i q_i.$$

Substituting $A^{1/2}q$ for q and $A^{-1/2}c$ for c with a symmetrical matrix A and then changing variables of integration, we easily obtain

$$\int \left|\det \frac{A}{2\pi}\right|^{1/2} \prod_{i=1}^{n} dq_i \, \exp\left(-\frac{1}{2}(q, Aq) + (c, q)\right) = \exp\left(\frac{1}{2}(c, A^{-1}c)\right).$$

$$(3.32)$$

Consider now the path

$$[q] = \{q(t) | \, t' \le t \le t''\}$$

as a vector with an infinite number of components $q(t)$ (an argument t playing the role of an index numbering these components) and analogously for the path

$$[c] = \{c(t) | \, t' \le t \le t''\}.$$

Then the last formula can be rewritten as follows:

$$\int d[q] \, \exp\left(-\frac{1}{2}([q], A[q]) + ([c], [q])\right) = \exp\left(\frac{1}{2}([c], A^{-1}[c])\right) \quad (3.33)$$

where the scalar product is defined for paths as

$$([c], [q]) = \int_{t'}^{t''} dt \, c(t)q(t), \quad (3.34)$$

A is a linear operator in the space of paths, and the measure in this space is formally defined by the formula

$$d[q] = \left|\det \frac{A \, dt}{2\pi}\right|^{1/2} \prod_{t=t'}^{t''} dq(t). \quad (3.35)$$

The latter formula should be understood as a receipt for the procedure of skeletonization. It is not difficult to see that the earlier defined path integral (3.12) is in accordance with this receipt up to a finite numerical factor. But the benefit of the formula (3.33) is that it is very easy to use this formula for operations with path integrals and to develop the perturbation theory.

To show this, let us take equations (3.33) and (3.34) as a formal definition of a *Gaussian path integral*. The concrete scheme of skeletonization expressed in equation (3.35) may be forgotten because in many cases there is no need to introduce the procedure of skeletonization explicitly.

Equation (3.33) can be used for evaluation of path integrals with different linear operators A including differential operators. For example, the choice

$$A = \frac{im}{\hbar}\left(\frac{d^2}{dt^2} + \omega^2\right), \quad c(t) = \frac{i}{\hbar}F(t) \qquad (3.36)$$

converts the integral (3.33) into the path integral for a driven harmonic oscillator (see the preceding section), in which the integral $\int \dot{q}^2\, dt$ is presented (with the help of integration by parts and up to a boundary term) in the form $-\int q\ddot{q}\, dt$. The formula (3.33) then gives for this integral the expression

$$\exp\left(-\frac{i}{2\hbar}([F], A^{-1}[F])\right).$$

This proves to coincide (up to a boundary term) with $\exp\left(\frac{i}{\hbar}S[q_{\text{class}}]\right)$. Thus the propagator (3.30) or (3.24) for a driven oscillator can be obtained with the help of equation (3.33) up to a numerical factor.

Remark 3 The difference in numerical factors arose from different definitions of path integral measures in the two schemes of evaluation. However, in many cases a numerical factor is not important. Moreover, a numerical factor can be found independently of the evaluation of a functional dependence of a path integral, as has been done in the preceding section for the oscillator path integral.

Taking derivatives of both sides of equation (3.32) with respect to components of the vector c shows that this is equivalent to introducing a product of the corresponding components of the vector q into an integrand. Therefore, the following formula is valid for a polynomial $\Phi(q)$:

$$\int \left|\det\frac{A}{2\pi}\right|^{1/2} \prod_{i=1}^{n} dq_i\, \Phi(q)\exp\left(-\frac{1}{2}(q, Aq) + (c, q)\right)$$
$$= \Phi\left(\frac{\partial}{\partial c}\right)\exp\left(\frac{1}{2}(c, A^{-1}c)\right).$$

Since any function can be approximated by a polynomial to arbitrary precision, this formula is valid in fact for any function $\Phi(q)$.

Generalizing this to the case of functions (paths), one has the useful formula

$$\int d[q] \, \Phi[q] \exp\left(-\frac{1}{2}([q], A[q]) + ([c], [q])\right)$$
$$= \Phi\left[\frac{\delta}{\delta c}\right] \exp\left(\frac{1}{2}([c], A^{-1}[c])\right) \tag{3.37}$$

where Φ is a functional and $\delta/\delta c(t)$ is a functional derivative. Taking $c \equiv 0$ in this formula, we obtain

$$\int d[q] \, \Phi[q] \exp\left(-\frac{1}{2}([q], A[q])\right)$$
$$= \Phi\left[\frac{\delta}{\delta c}\right] \exp\left(\frac{1}{2}([c], A^{-1}[c])\right)\Bigg|_{c \equiv 0}. \tag{3.38}$$

This equation allows one to integrate any functional with the Gaussian measure. The result will be obtained in the form of a series, giving in fact a perturbation expansion for such an integral.

Remark 4 The Gaussian integral (3.15) converges for any complex parameter a satisfying $\mathrm{Re}\, a < 0$. Correspondingly, all its generalizations, for example (3.32), (3.33), (3.37) and (3.38), have meaning for an operator A with a positive Hermitian part. For a purely anti-Hermitian operator A as in equation (3.36) one should, for a correct definition, introduce a positive Hermitian part and then take the limit when this part tends to zero.

4

Continuous Measurement and Evolution of the Measured System

The main ideas leading to the path-integral approach in the theory of continuous quantum measurements have been formulated in Chapter 1. In this chapter we shall give a more accurate formulation of this approach, including a description of the evolution of the measured system and the generalized unitarity condition for this evolution. It is argued that evolution under continuous measurement has both quantum and classical features. This is why it is convenient in the framework of this formalism to trace the transition from the quantum picture of evolution to the classical one.

The method of the effective Lagrangian (effective action) is considered in section 4.2 as a simple way of evaluating the measurement amplitude (the main object of the present approach) instead of direct calculation of a path integral.

Sections 4.3 and 4.4 deal with the evolution of the system subject to continuous measurement. We can see from them that obtaining information about the behaviour of the system influences this behaviour. In section 4.3 this will be demonstrated in general form, and in section 4.4 we consider the specific example of a free particle scattered because its position is measured (monitored).

Sections 4.3 and 4.4 may be skipped in a first reading.

4.1 THE MEASUREMENT AMPLITUDE

The quantum theory of measurements has been discussed actively in recent years not only because of its conceptual interest (which has always been discussed from the very first days of quantum mechanics) but also for practical purposes. Indeed, the precision of measurements has nowadays become so high that quantum effects are important in many cases. This is true for optical measurements and for high-precision mechanical measurements such as those performed in gravitational-wave antennas. It is in connection with the theory of gravitational-wave detectors that practical

48

aspects of quantum measurements have for the first time been investigated seriously (Braginsky 1967).

In practice, measurements are usually rather complicated and in most cases cannot be considered to be instantaneous. The finite duration of measurements has given the theory of quantum continuous measurements a real application. The path-integral approach we are discussing here turned out to be one of the most effective approaches in this theory. The idea of this approach was proposed in a short remark in the paper of Feynman (1948). The present author developed this approach without being aware of Feynman's proposal (Mensky 1979a, b, 1983a). Let us mention, however, that attempts (not successful, in our opinion) to elaborate Feynman's idea were made in some other papers (see, for example, Bloch and Burba 1974 and references therein). Results close to those of the present author but in the framework of another method were obtained by Khalili (1981) and Barchielli *et al* (1982). They constructed the model of continuous measurement as a series of instantaneous measurements with the interval between them tending to zero. The theory of continuous quantum measurements has since been developed further by Caves (1986, 1987) with the help of path integrals.

The path-integral approach to continuous quantum measurements can be justified in two different ways.

- One is the derivation of the path-integral description of measurement in the framework of the limiting process from sequential to continuous measurements. In this case the conventional (von Neumann's) description is applied to instantaneous measurements included in the sequential one. This derivation was considered briefly at the end of Chapter 2. A more detailed derivation can be found in Mensky (1983a).

- Another way consists in the derivation of the corresponding probability amplitudes directly from the physical meaning of the partial amplitudes that form the Feynman path integral. This approach has been applied in Chapter 1, and more details are given by Mensky (1983a).

The second of these two ways is more natural and leads to a wider applicability of the approach, though the first way may be desirable for making it quite convincing. Here we will not repeat the argument leading to the path-integral approach (see Chapter 1). Instead, we shall give its final formulation, considering not only the probabilities of different outputs of a continuous measurement but also the evolution of the system undergoing such a measurement.

Let a quantum system be characterized by the Lagrangian

$$L = L(q, \dot{q}, t)$$

and by the action functional

$$S[q] = \int_{t'}^{t''} L(q, \dot{q}, t)\, \mathrm{d}t$$

where q is a point of (generally multidimensional) configuration space. Then the probability amplitude for the system to transit from the point q' at the instant t' to the point q'' at the instant t'', or the *propagator* of the system, can be expressed by the *Feynman path integral* [1]

$$U(t'', q'' | t', q') = \int_{q'}^{q''} \mathrm{d}[q]\, e^{\frac{i}{\hbar} S[q]} \tag{4.1}$$

over all paths

$$[q] = \{q(t) | t' \le t \le t''\}$$

connecting the above-mentioned points:[2]

$$q(t') = q', \quad q(t'') = q''.$$

This is valid for the case when no measurement is performed in the time interval $[t', t'']$ so that no information is available about the specific path $[q]$ chosen by the system for the transition (this is why integration over all of them is necessary).

If a *continuous measurement*[3] is performed in the time interval $[t', t'']$ then some information about this path is available, at least in principle. Let this information be expressed by the (positive-valued) functional w_α,

$$0 \le w_\alpha \le 1.$$

This means that knowlege of the measurement output α allows one to estimate as probable those paths $[q]$ for which $w_\alpha[q]$ is close to unity. On the contrary, the paths $[q]$ for which $w_\alpha[q]$ is close to zero should be considered to be improbable in the light of information supplied by the measurement output α. In fact the result (output) α of the continuous measurement can be completely characterized by specifying the corresponding functional w_α.

---

[1] The definition and methods of evaluation of path integrals are given in Chapter 3. However, they are not necessary for understanding the main ideas and simplest applications of the present approach.

[2] We introduced these points explicitly into the notation for a path integral.

[3] We investigate continuous measurements here, but in fact an instantaneous measurement can be considered as a special case of continuous one, giving information only about the behaviour of a path $[q]$ in the vicinity of a fixed instant t.

If a continuous measurement of the given type is performed to give the output α, then the probability amplitude for the system to transit from the point (q', t') to (q'', t'') is equal to the restricted (weighted) path integral

$$U_\alpha(t'', q'' | t', q') = \int_{q'}^{q''} d[q]\, w_\alpha[q]\, e^{\frac{i}{\hbar} S[q]}. \tag{4.2}$$

In a special case when the functional w_α takes only the value 1 or 0, the integral (4.2) can be considered to be the Feynman path integral restricted to the set of paths $[q]$ for which $w_\alpha[q] = 1$. It is convenient to denote this set by the same letter α as the corresponding measurement output. Then the integral (4.2) takes the form

$$U_\alpha(t'', q'' | t', q') = \int_\alpha d[q]\, e^{\frac{i}{\hbar} S[q]}. \tag{4.3}$$

This formula has an even more direct interpretation than equation (4.2): integration goes only over those paths which are compatible with the measurement output. Position monitoring as a typical example of continuous measurement will be considered in section 4.2 below.

In a general case the restricted integral (4.3) can be considered to be an approximate expression for the precise weighted integral (4.2), if one defines α as the set of paths $[q]$ such that $w_\alpha[q]$ is sufficiently close to unity (say, greater than e^{-1}). Let us call the probability amplitude (4.2) the *measurement amplitude*. It is intuitively clear that the measurement amplitude allows one to roughly estimate the probabilities of different measurement outputs. The output α_{class} for which $|U_\alpha|^2$ is maximum is the most probable.[4] The outputs α for which $|U_\alpha|^2$ is close to the maximum value $|U_{\alpha_{class}}|^2$ are also sufficiently probable. Those α for which $|U_\alpha|^2$ is much less are practically improbable.

This rough estimation turns out to be sufficient for many practical goals and will be in fact taken later as the basis of a convenient qualitative result called the action uncertainty principle. However, for a more precise formulation the corresponding probability measure should be defined. This will be done in section 4.3.

However, before this we shall consider in section 4.2 an important question about the choice of the weight functional w_α and the method of evaluating the amplitudes U_α.

---

[4] We shall see later that this output corresponds to the prediction of classical theory. This is a consequence of the fact that the action $S[q]$ changes slowly in the vicinity of the classical trajectory $[q_{class}]$, and thus the amplitudes $\exp\left(\frac{i}{\hbar} S[q]\right)$ interfere constructively in this vicinity.

4.2 EFFECTIVE LAGRANGIAN

The measurement amplitude $U_\alpha(t'', q''|t', q')$ is the key object in the path-integral approach to continuous quantum measurements. It is defined by equation (4.2) or equation (4.3) as a restricted or weighted path integral. It is natural to ask how the measurement amplitudes can be evaluated.

The direct answer to this question is to refer to different methods of evaluating path integrals. These methods have been reviewed briefly in section 3.3 of Chapter 3, and evaluation of Gaussian path integrals has been considered in more detail in sections 3.3 and 3.4 of Chapter 3. There is, however, an indirect way of finding the measurement amplitude (this method has also been mentioned in section 3.3), leading in many cases to a simpler procedure. We mean solution of the Schrödinger equation. This way is more customary for many people and is more efficient in the case of nonlinear systems. Let us consider this method in some detail.

The starting point is the fact that the Feynman propagator (4.1) satisfies the Schrödinger equation

$$i\hbar \frac{\partial}{\partial t''} U(t'', q''|t', q') = \hat{H}'' U(t'', q''|t', q') \tag{4.4}$$

and the initial condition

$$U(t', q''|t', q') = \delta(q'' - q'). \tag{4.5}$$

Here $\hat{H}$ is the quantum Hamiltonian of the system under consideration. This means that it can be obtained by the replacement $p \to -i\hbar \frac{\partial}{\partial q}$ from the classical Hamiltonian

$$H(q, p, t) = p\dot{q} - L(q, \dot{q}, t), \qquad p = \frac{\partial L}{\partial \dot{q}}$$

(the multidimensional generalization of these formulas is straightforward). The primes of the operator $\hat{H}''$ in equation (4.4) indicate that it acts on the first argument q'' of the propagator.

Instead of evaluating the integral (4.1) to find the propagator, one can therefore solve equation (4.4) with the initial condition (4.5). In most cases this procedure is simpler than direct evaluation of a path integral.

Turning now to the measurement amplitude (4.2), we see that the same method can be applied to its evaluation provided that the weight functional

$w_\alpha[q]$ can be presented in the form of an exponential of a time integral:[5]

$$w_\alpha[q] = \exp\left(-\gamma \int_{t'}^{t''} K_\alpha(q, \dot{q}, t)\, dt\right). \tag{4.6}$$

Then the path integral (4.2) can be written in a complete analogy with the Feynman integral,

$$U_\alpha(t'', q''|t', q') = \int_{q'}^{q''} d[q]\, e^{\frac{i}{\hbar} S_\alpha[q]} \tag{4.7}$$

but with the *effective action*

$$S_\alpha[q] = S[q] + i\hbar\gamma \int_{t'}^{t''} K_\alpha(q, \dot{q}, t)\, dt.$$

This can be rewritten with the help of an effective Lagrangian:

$$\begin{aligned} S_\alpha[q] &= \int_{t'}^{t''} L_\alpha(q, \dot{q}, t)\, dt \\ L_\alpha &= L + i\hbar\gamma\, K_\alpha. \end{aligned}$$

To see that this definition is not artificial let us consider a typical example of continuous measurement called *monitoring of position* or *measurement of a path*. This measurement means estimating the position $q(t)$ at each instant t with some final error Δa.[6] If the estimation is $a(t)$, this means that the actual position $q(t)$ differs from its estimation $a(t)$ by no more than the value Δa.

This verbal description of the measurement can be made mathematically precise with the help of the inequality

$$|q(t) - a(t)| \leq \Delta a. \tag{4.8}$$

Then the result of the measurement,

$$[a] = \{a(t)|\, t' \leq t \leq t''\},$$

defines the set of paths $I_{[a]}$, which can be described as a corridor of width $2\Delta a$ with the path $[a]$ lying in the middle (figure 4.1). The measurement amplitude $U_{[a]}$ in this case can be evaluated as a path integral (4.3) over the corridor $\alpha = I_{[a]}$. This procedure has been performed by Mensky (1979a).

---

[5] The factor γ here is factored out for convenience because it can emerge naturally in an application as a characteristic of the measurement precision.

[6] We consider the same error Δa for all instants t though an arbitrary dependence $\Delta a(t)$ of the error on time can be treated quite analogously.

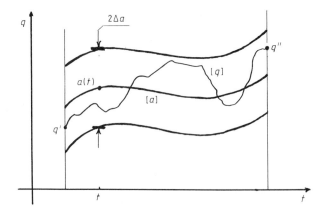

Figure 4.1: A corridor $\alpha = I_{[a]}$ describing the monitoring of position (measurement of a path).

Remark 1 This analysis justifies another name for position monitoring. It may be called *measurement of a path*. The result of this measurement is naturally denoted by a path [a]. However, it is actually expressed not by a single path but by a corridor of finite width around this path.

This definition, though quite transparent, is neither realistic nor simple to deal with. It corresponds to a measurement of the position $q(t)$ at each instant t with the help of a measuring device having a rectangular characteristic. Given the measurement output $a(t)$, such a device guarantees that the actual position $q(t)$ lies in the interval

$$[a(t) - \Delta a, \, a(t) + \Delta a]$$

with the same probability density for each point of the interval, but certainly cannot lie outside the interval. Of course, real measuring devices give information of a different type, with some smooth probability distribution around the point $a(t)$. The description of measurement with the help of such a device has been considered in section 2.3 of Chapter 2. We will not repeat this consideration here. Instead, we describe the resulting measurement directly in terms of paths, which proves to be much simpler.

The output of the measurement under consideration is some path [a]. This path is of course the best estimation of the actual path [q] that can be given on the basis of the measurement output. However, paths in some vicinity of [a] are actually equally probable (or only slightly less probable than [a]). For a device with a rectangular characteristic this vicinity can be described by the inequality (4.8), i.e. as a corridor with sharp edges. For a corridor with unsharp edges we can accept a definition based upon the square averaged deviation $\rho[q - a]$ of $[q]$ from $[a]$:

$$\rho^2[q - a] = \overline{(q - a)^2} = \frac{1}{t'' - t'} \int_{t'}^{t''} \left(q(t) - a(t)\right)^2 \mathrm{d}t.$$

If the monitoring of position has given the output $[a]$, then we know that $\rho[q - a]$ is not large, no greater than Δa. This means that the weight functional $w_{[a]}[q]$ in (4.2) should be taken close to unity for $\rho \lesssim \Delta a$ and close to zero for $\rho \gg \Delta a$. In most cases it is sufficiently realistic to choose the weight functional in the Gaussian form:

$$w_{[a]}[q] = \exp \left(-\frac{2\rho^2[q - a]}{\Delta a^2}\right). \tag{4.9}$$

Gaussian weights leading to Gaussian path integrals will be used in most of the calculations in this book, and therefore their applicability should be justified. The weight functional w_α describes information resulting from the measurement output α. Therefore the choice of the functional depends on what type of measurement we wish to investigate. Strictly speaking, one should analyse some class of real measurement devices or real measurement situations and *deduce* the form of the weight functional from this analysis. Results of all calculations in the framework of the path-integral approach depend on the chosen form of the functional.

However, one may suppose that a small change in weight functional cannot change the results of the calculation radically. At least for all estimations up to the order of magnitude (which is the main goal of most calculations in this book) details in the definition of the functional can be expected to be unimportant. In this situation the functional most convenient for calculation may be chosen.

Practice shows that this is actually the case. In the very first paper on coordinate monitoring (Mensky 1979a) the author described this type of measurement using the functional

$$w'_{[a]}[q] = \begin{cases} 1 & \text{for } |q(t) - a(t)| \leq \Delta a \quad (t' \leq t \leq t'') \\ 0 & \text{otherwise.} \end{cases} \tag{4.10}$$

This choice corresponds to a rectangular characteristic of the measuring device. Calculations with this functional were very complicated. In the next paper (Mensky 1979b) a Gaussian functional (4.9) was chosen that was simpler to use and actually more realistic. The results of the two calculations coincided up to the order of magnitude, the precision sufficient for analysis of real measurements.

Of course such an example does not prove anything, and the problem of choice of the weight functional deserves further analysis, but we shall use Gaussian functionals in most considerations.[7]

---

[7] It will be shown in Chapter 10, however, that this choice of a functional is not quite appropriate from the point of view of the semigroup describing time evolution.

The Gaussian weight (4.9) for description of coordinate monitoring (which can also be called measurement of a path) leads to an effective Lagrangian with a quadratic imaginary term:

$$K_{[a]}(q, \dot{q}, t) = \left(q - a(t)\right)^2$$

if we denote

$$\gamma = \frac{2}{\Delta a^2(t'' - t')}.$$

Remark 2 It is important to remember that $t'' - t'$ here should be considered to be constant, an effective Lagrangian depending on time only through the dependence of $K_{[a]}$ on $a(t)$.

4.3 EVOLUTION OF A QUANTUM SYSTEM SUBJECT TO CONTINUOUS MEASUREMENT

In section 4.1 we discussed briefly the application of the measurement amplitude for estimation of probabilities of different measurement outputs. However, the same amplitude can be used to describe the evolution of the system subject to the measurement.

Instead of a two-point function $U(t'', q''|t', q')$ defined by equation (4.1) one can introduce the corresponding operator $U_{t''t'}$ in such a way that its matrix elements between position eigenvectors are equal to the two-point function under consideration:

$$U(t'', q''|t', q') = \langle q''|U_{t''t'}|q'\rangle.$$

Remark 3 The position eigenvector $|q'\rangle$ is an eigenvector of the position operator: $\hat{q}|q'\rangle = q'|q'\rangle$. Such eigenvectors are not proper vectors of Hilbert space (they are unnormalizable) and are usually normalized by a delta function: $\langle q'|q''\rangle = \delta(q' - q'')$. They form a complete set of vectors in the sense that $\int |q'\rangle\langle q'| \, dq' = 1$. The value of the wavefunction $\psi(q')$ describing the state $|\psi\rangle$ (the latter being understood as a vector of the Hilbert space of states) is then equal to the scalar product of the position eigenvector and the state vector, $\psi(q') = \langle q'|\psi\rangle$. One can say that the function $U(t'', q''|t', q')$ is a kernel of the operator $U_{t''t'}$, the latter being considered as an integral operator:

$$(U_{t''t'}\psi)(q'') = \int U(t'', q''|t', q') \, \psi(q') \, dq'.$$

If the two-point function $U(t'', q''|t', q')$ is equal to the propagator (4.1), then the corresponding operator $U_{t''t'}$ is just an evolution operator. It describes the evolution of the system from the state $|\psi_{t'}\rangle$ at the instant t' into the state $|\psi_{t''}\rangle$ at the instant t'' as follows:

$$|\psi_{t''}\rangle = U_{t''t'}|\psi_{t'}\rangle.$$

Knowing the quantum Hamiltonian $\hat{H}_t = H(\hat{q}, \hat{p}, t)$ of the system, one can express the evolution operator as

$$U_{t''t'} = \exp\left(-\frac{i}{\hbar}\int_{t'}^{t''}\hat{H}_t\,dt\right).$$

This operator is unitary in the sense that the equation

$$U_{t''t'}^\dagger U_{t''t'} = 1$$

is valid.[8]

Let us now suppose that a continuous measurement is performed in the system during the interval $[t', t'']$, resulting in the output α. Then the role of an evolution operator is played (Mensky 1988a) by the operator $U_{t''t'}^\alpha$ corresponding to the measurement amplitude (4.2):

$$U_\alpha(t'', q''|t', q') = \langle q''|U_{t''t'}^\alpha|q'\rangle.$$

The evolution of the system undergoing the measurement (with the given output) is then described by the formula

$$|\psi_{t''}^\alpha\rangle = U_{t''t'}^\alpha|\psi_{t'}\rangle. \tag{4.11}$$

Of course, the result of this evolution depends on the output α of the measurement. In this sense the operator $U_{t''t'}^\alpha$ may be called a *partial evolution operator*. A complete picture of the evolution of the system for all possible measurement outputs α is described by the set of partial evolution operators $U_{t''t'}^\alpha$ for all α.

The formula (4.11) for evolution of the system state is valid if the output α of the measurement is known. However, it is important to have a description of the evolution that does not depend on our information about the result of the measurement. It is important in two different situations:

- When we want to analyse the evolution of the system after the measurement is over but we do not know its result (output): *a posteriori analysis*.

---

[8] An Hermitian conjugation is defined by $\langle\varphi|A^\dagger|\psi\rangle = \langle\psi|A|\varphi\rangle^*$ so that a unitary operator (satisfying the condition $U^\dagger U = 1$) conserves the scalar product: $\langle\varphi'|\psi'\rangle = \langle\varphi|\psi\rangle$ for $|\psi'\rangle = U|\psi\rangle$, $|\varphi'\rangle = U|\varphi\rangle$.

- When we want to analyse the possible evolution of the system before the measurement is performed and therefore no information about its output is available in principle: *a priori analysis*.

In both cases the term '*non-selective measurement*' or 'non-selective analysis' may be used. For analysis of non-selective measurement (or, more precisely, for non-selective analysis of a measurement) some sort of summation over all possible measurement outputs α should be performed so that all of them are taken into account.

It is high time to recall the different types of summation over alternatives discussed in section 1.2 of Chapter 1. We argued there (following Feynman and Hibbs (1965)) that there are two different types of alternative and two different types of summation:

- classic, or incompatible, alternatives characteristic of classical physics with summation of probabilities of these alternatives,

- and quantum, or interfering, alternatives characteristic of quantum physics with summation of their probability amplitudes.

We saw that classical alternatives can arise even in the framework of quantum theory provided that they are controlled by some processes which can be (in a correct approximation) described as classical. Different outputs of a measurement are always classical alternatives, and if we want to sum over different measurement outputs, we should add not amplitudes but probabilities.

If the measurement output is known to be α, then the state $|\psi_{t'}\rangle$ converts after evolution into $|\psi_{t''}^{\alpha}\rangle$, in correspondence with equation (4.11). One could suppose that the result of evolution should be $\sum_{\alpha} |\psi_{t''}^{\alpha}\rangle$ in the case where the measurement output is not known (non-selective situation). However, this would mean summation of amplitudes, which is wrong for classical alternatives α. Instead of this, a transition from state vectors to density matrices should be performed before summation.

For such a transition let us begin from the pure state $|\psi_{t'}\rangle$. Instead of a state vector the same state may be described by the density matrix

$$\rho_{t'} = |\psi_{t'}\rangle\langle\psi_{t'}|.$$

The evolution of this state under the condition that the measurement results in an output α leads to the state described by the density matrix

$$\rho_{t''}^{\alpha} = |\psi_{t''}^{\alpha}\rangle\langle\psi_{t''}^{\alpha}|.$$

This gives, due to equation (4.11), the law of evolution in terms of density matrices:

$$\rho_{t''}^{\alpha} = U_{t''t'}^{\alpha} \, \rho_{t'} \, (U_{t''t'}^{\alpha})^{\dagger} . \tag{4.12}$$

Though equation (4.12) is derived for a pure initial state, it turns out to be valid in the case of a mixed state as well. Indeed, each density matrix $\rho_{t'}$, even describing a mixed state, can be presented in the form

$$\rho_{t'} = \sum_i p_i \, |\psi_{t'}^{(i)}\rangle\langle\psi_{t'}^{(i)}|$$

with positive numbers p_i satisfying the condition

$$\sum_i p_i = 1$$

equivalent to the normalization condition for a density matrix:

$$\operatorname{tr} \rho_{t'} = 1. \tag{4.13}$$

These numbers have the sense of probabilities, and the mixed state $\rho_{t'}$ may be equivalently described by claiming that one of the pure states $|\psi_{t'}^i\rangle$ takes place, the ith state with probability p_i. The evolution of the mixed state can therefore be described as the evolution of a set of pure states. This leads to a new mixed state with the same set of probabilities p_i but with another set of pure states:

$$\rho_{t''}^\alpha = \sum_i p_i \, |\psi_{t''}^{(i)\alpha}\rangle\langle\psi_{t''}^{(i)\alpha}|$$

where

$$|\psi_{t''}^{(i)\alpha}\rangle = U_{t''t'}^\alpha \, |\psi_{t'}^{(i)}\rangle$$

is obtained according to the evolution law (4.11). It is evident that the resulting density matrix $\rho_{t''}^\alpha$ may be expressed in the form (4.12) even for a mixed initial state.

Now it is easy to go over to the non-selective case when the measurement output is not known. It is sufficient to sum equation (4.12) over different measurement outputs α. The resulting formula is

$$\rho_{t''} = \sum_\alpha \rho_{t''}^\alpha = \sum_\alpha U_{t''t'}^\alpha \, \rho_{t'} \, (U_{t''t'}^\alpha)^\dagger. \tag{4.14}$$

Given a normalized (in the sense of equation (4.13)) density matrix $\rho_{t'}$ we obtain $\rho_{t''}$ as a result of evolution, and to make the description correct, this density matrix should be normalized too:

$$\operatorname{tr} \rho_{t''} = 1.$$

It can be proved that this is the case (for an arbitrary normalized initial density matrix) if and only if the following condition is fulfilled:

$$\sum_\alpha (U_{t''t'}^\alpha)^\dagger \, U_{t''t'}^\alpha = 1. \tag{4.15}$$

This condition on the set of all partial evolution operators plays the role of a *generalized unitarity condition* for a system under continuous measurement. The physical content of this condition is the same as for a conventional unitarity condition: conservation of probability.[9]

In the last formulas we used summation over the measurement outputs α. However, the set of all outputs is of course continuous (not discrete). The summation should be therefore defined in a more accurate way, rather as integration with a measure $\mu(\alpha)$. The formulas (4.14) and (4.15) then have the form

$$\rho_{t''} = \int d\mu(\alpha)\, U_{t''t'}^{\alpha}\, \rho_{t'}\, (U_{t''t'}^{\alpha})^{\dagger} \tag{4.16}$$

$$\int d\mu(\alpha)\, (U_{t''t'}^{\alpha})^{\dagger}\, U_{t''t'}^{\alpha} = 1. \tag{4.17}$$

It should be emphasized that the partial density matrices $\rho_{t''}^{\alpha}$ corresponding to separate measurement outputs (and therefore to separate evolution channels) are not normalized. Instead, the value $\operatorname{tr}\rho_{t''}^{\alpha}$ has the sense of a probability density for the measurement to result in the output α. The probability for the measurement to give an output in an infinitesimal interval $[\alpha, \alpha + d\alpha]$ is equal to

$$\operatorname{prob}(\alpha, d\alpha) = \operatorname{tr}\rho_{t''}^{\alpha}\, d\mu(\alpha). \tag{4.18}$$

The probability for the measurement to result in an output from the set A of possible outputs can be found by integration:

$$\operatorname{prob}(\alpha \in A) = \int_{A} \operatorname{tr}\rho_{t''}^{\alpha}\, d\mu(\alpha). \tag{4.19}$$

The generalized unitarity condition (4.17) ensures that the probability corresponding to the complete set of measurement outputs is equal to unity.

4.4 SCATTERING BY THE MEASURING MEDIUM

We saw in section 4.3 that continuous measurement changes the evolution of the quantum system undergoing the measurement. The evolution of the measured system can be described by a set of partial evolution operators $U_{t''t'}^{\alpha}$ corresponding to all possible outputs α of the measurement. Starting from the density matrix $\rho_{t'}$ at the instant t', the system ends with the density matrix $\rho_{t''}$ determined by equation (4.16), if the measurement output is not known (non-selective situation). Let us consider this evolution law in more detail.

---

[9] Larsen (1986) investigated some mathematical aspects of the evolution law (4.14).

Since the time interval $[t', t'']$ will be fixed for our purposes, it will be more convenient to put $t' = 0$ and denote $t'' = T$ and $U_{T0}^\alpha = U_\alpha$. Then instead of equation (4.16) one has for the evolution law

$$\rho_T = \int d\mu(\alpha) \, U_\alpha \, \rho_0 U_\alpha^\dagger. \tag{4.20}$$

The generalized unitarity condition equation (4.17) is then converted into

$$\int d\mu(\alpha) \, U_\alpha^\dagger U_\alpha = 1. \tag{4.21}$$

The matrix elements of the operator U_α are given by the formula (4.2)

$$U_\alpha(q', q) = \langle q' | U_\alpha | q \rangle = \int_q^{q'} d[q] \, w_\alpha[q] \, e^{\frac{i}{\hbar} S[q]}. \tag{4.22}$$

where integration goes over paths starting in q and terminating in q'.

Now, presenting density matrices with the help of their coordinate representation, i.e. by the matrix elements

$$\rho_t(q_1, q_2) = \langle q_1 | \rho_t | q_2 \rangle$$

we have for the evolution law the following formula:

$$\rho_T(q_1', q_2') = \int dq_1 \int dq_2 \, U(q_1', q_1 | q_2', q_2) \rho_0(q_1, q_2)$$

where the following notation is introduced:

$$U(q_1', q_1 | q_2', q_2) = \int d\mu(\alpha) \, U_\alpha(q_1', q_1) U_\alpha^*(q_2', q_2). \tag{4.23}$$

The generalized unitarity condition can be written with the help of this notation as follows:

$$\int dq_1' \, U(q_1', q_1 | q_1', q_2) = \delta(q_1 - q_2).$$

The four-point function (4.23) plays role of a *superpropagator* (i.e. evolution operator for a density matrix), but for a system subject to continuous measurement. Making use of the formula (4.22), we have for a superpropagator

$$U(q_1', q_1 | q_2', q_2) = \int_{q_1}^{q_1'} d[q_1] \int_{q_2}^{q_2'} d[q_2] \, w([q_1], [q_2]) \, \exp\left(\frac{i}{\hbar}(S[q_1] - S[q_2])\right) \tag{4.24}$$

where

$$w([q_1], [q_2]) = \int d\mu(\alpha)\, w_\alpha[q_1] w_\alpha[q_2].$$

Let us illustrate these general statements in the case of the evolution of a free particle having its position monitored.

The superpropagator (4.24) has been evaluated for a class of continuous measurements (called spectral measurements) of a harmonic oscillator by Mensky (1988a). (We shall consider this type of measurement in section 5.4 of Chapter 5.) Monitoring of position (4.9) is in fact a special case of spectral measurements, while a (one-dimensional) free particle

$$L = \frac{1}{2} m \dot{q}^2$$

is a special case of an oscillator (corresponding to zero frequency, $\omega = 0$). According to Mensky (1988a), the calculation of a path integral (4.22) and of an integral (4.23)[10] gives for the superpropagator in this case

$$U(q_1', q_1 | q_2', q_2) = N \exp\left(\frac{im}{2\hbar T} [(q_1' - q_1)^2 - (q_2' - q_2)^2] \right.$$
$$\left. - \frac{1}{3\Delta a^2} [(q_1 - q_2)^2 + (q_1' - q_2')^2 + (q_1 - q_2)(q_1' - q_2')] \right).$$

This of course can be trivially generalized to the case of a three-dimensional free particle

$$L = \frac{1}{2} m \dot{\boldsymbol{r}}^2 = \frac{1}{2} m (\dot{x}^2 + \dot{y}^2 + \dot{z}^2)$$

and monitoring of the position of this particle giving the output

$$[\boldsymbol{a}] = \{\boldsymbol{a}(t) \,|\, 0 \le t \le T\}.$$

The Gaussian weight functional describing monitoring should now be taken as follows:

$$w_{[a]}[\boldsymbol{r}] = \exp\left(-\frac{2}{T\Delta a^2} \int_0^T (\boldsymbol{r}(t) - \boldsymbol{a}(t))^2 \, dt \right). \tag{4.25}$$

This leads to a superpropagator of the form

$$U(\boldsymbol{r}_1', \boldsymbol{r}_1 | \boldsymbol{r}_2', \boldsymbol{r}_2) = N \exp\left(\frac{im}{2\hbar T} [(\boldsymbol{r}_1' - \boldsymbol{r}_1)^2 - (\boldsymbol{r}_2' - \boldsymbol{r}_2)^2] \right.$$
$$\left. - \frac{1}{3\Delta a^2} [(\boldsymbol{r}_1 - \boldsymbol{r}_2)^2 + (\boldsymbol{r}_1' - \boldsymbol{r}_2')^2 + (\boldsymbol{r}_1 - \boldsymbol{r}_2)(\boldsymbol{r}_1' - \boldsymbol{r}_2')] \right).$$

[10] An appropriate choice of the measure in this integral can be shown (Mensky 1988a) to be $d\mu(\alpha) = d[a] = \prod_{t=0}^{T} da(t)$.

With this superpropagator, the evolution of a free particle having its position monitored (with accuracy Δa) is described by

$$\rho_T(\boldsymbol{r}_1', \boldsymbol{r}_2') = \int d\boldsymbol{r}_1 \int d\boldsymbol{r}_2 \, U(\boldsymbol{r}_1', \boldsymbol{r}_1 | \boldsymbol{r}_2', \boldsymbol{r}_2)\rho_0(\boldsymbol{r}_1, \boldsymbol{r}_2). \qquad (4.26)$$

We shall calculate the average

$$\langle \boldsymbol{p} | \rho_T | \boldsymbol{p} \rangle$$

of the density matrix ρ_T in the state $|\boldsymbol{p}\rangle$ of a definite linear momentum, corresponding to the wavefunction

$$\psi_{\boldsymbol{p}}(\boldsymbol{r}) = \langle \boldsymbol{r} | \boldsymbol{p} \rangle = N_1 \, \exp\left(\frac{i}{\hbar} \boldsymbol{p}\boldsymbol{r}\right).$$

This average gives a distribution over different values of linear momentum of a particle at the instant T. Then we can compare the distributions corresponding to different errors Δa of position monitoring. Specifically, we can compare the distribution of linear momentum under monitoring with accuracy Δa with the case $\Delta a = \infty$ when no monitoring is actually performed.

Gaussian integration (according to the formulas of section 3.4 of Chapter 3) gives easily

$$\langle \boldsymbol{p} | \rho_T | \boldsymbol{p} \rangle = \int d\boldsymbol{r}_1 \int d\boldsymbol{r}_2 \, W_{\boldsymbol{p}}(\boldsymbol{r}_1, \boldsymbol{r}_2)\rho_0(\boldsymbol{r}_1, \boldsymbol{r}_2). \qquad (4.27)$$

where

$$W_{\boldsymbol{p}}(\boldsymbol{r}_1, \boldsymbol{r}_2) = N_2 \exp\left(-\frac{i}{\hbar} \boldsymbol{p}(\boldsymbol{r}_1 - \boldsymbol{r}_2) - \frac{2}{\Delta a^2}(\boldsymbol{r}_1 - \boldsymbol{r}_2)^2\right).$$

Let an initial state of a particle be

$$\psi_0(\boldsymbol{r}) = N_3 \exp\left(\frac{i}{\hbar} \boldsymbol{p}_0 \boldsymbol{r} - \frac{1}{2l^2} r^2\right).$$

This state describes a particle located at the origin with uncertainty $\Delta q_0 = l/\sqrt{2}$ and having the linear momentum $\boldsymbol{p}_0$ with uncertainty (in each component)

$$\Delta p_0 = \frac{\hbar}{\sqrt{2}l}.$$

The density matrix for this state is

$$\rho_0(\boldsymbol{r}_1, \boldsymbol{r}_2) = |N_3|^2 \exp\left(-\frac{i}{\hbar} \boldsymbol{p}_0(\boldsymbol{r}_1 - \boldsymbol{r}_2) - \frac{1}{2l^2}(r_1^2 + r_2^2)\right).$$

Substituting this density matrix into equation (4.27) and evaluating a Gaussian integral, one has

$$\langle p|\rho_T|p\rangle = N_4 \exp\left(-\frac{(p-p_0)^2}{2\left(\frac{\hbar^2}{l^2}+\frac{4\hbar^2}{\Delta a^2}\right)}\right).$$

Thus the variance of linear momentum at the instant T (i.e. after continuous measurement) is equal to

$$\Delta p = \sqrt{\Delta p_0^2 + \frac{2\hbar^2}{\Delta a^2}}. \tag{4.28}$$

The corresponding formula for a particle propagating without any measurement can be obtained by taking $\Delta a = \infty$ (measurement with infinite error means no measurement at all). It is natural that the linear momentum variance of a free particle does not change with time because of conservation of linear momentum. The influence of position monitoring consists in widening of the momentum variance according to the formula (4.28).

We found only one characteristic of the state ρ_T resulting at the instant T after evolution with the influence of continuous measurement. This state can be found in explicit form with the help of the formula (4.26). However, even one characteristic of this state (4.28) shows that measurement has an influence on the evolution of a particle. One can say that a particle undergoes *scattering as a result of continuous measurement*.

In the case considered here the influence of measurement consists in widening of the linear momentum distribution. This widening can be easily understood from the point of view of the uncertainty principle. As a result of continuous measurement the position of a particle is known with the error Δa. If this error is less than an initial indeterminacy in position, $\Delta a \ll \Delta q_0$, then, according to the uncertainty principle, the resulting indeterminacy of linear momentum will become $\Delta p \simeq \sqrt{2}\hbar/\Delta a$ instead of $\Delta p_0 = \hbar/2\Delta q_0$.

In discussing measurement (specifically position monitoring) we do not necessarily mean measurement arranged on purpose. It is important only that information about our system is left in some outside system. The particle might propagate through a specific medium with properties such that propagation of a particle in that medium leaves information about its state (specifically its position) at each instant with some accuracy. An evident example is a bubble or Wilson camera. However, any medium can in fact be considered to be a measuring medium. The question is only about the type of information this measurement gives and about the accuracy of the measurement.

Remark 4 The Gaussian corridor (4.9), (4.25) can be used to describe position monitoring only for a fixed time interval T. For consideration of continuous time evolution of the state (i.e. for comparing this state at different instants) the rectangular-characteristic approximation of the instant measurement (see section 2.3.3 of Chapter 2) is more appropriate. We shall discuss this later, when considering the group-theoretical features of continuous measurements (see section 10.3 of Chapter 10).

5

Continuous Measurements of Oscillators

In the preceding chapter we formulated a general method for considering continuous quantum measurements on the basis of path integrals. Here we shall apply this method to the simplest (but very important) quantum system: an oscillator. Two types of continuous measurement will be investigated for this system: position monitoring and spectral measurement (i.e. measuring some frequency components of a path).

Many real systems may be treated approximately as harmonic oscillators or systems of harmonic oscillators. This is why the results of the present chapter in fact have a wide area of application. We shall apply them to a system of practical interest, namely the gravitational-wave antenna of Weber type.

In section 5.5 we shall apply the method of effective Lagrangian (Hamiltonian) in combination with numerical computations to nonlinear oscillators. This shows how one can develop the method of computer calculation in the present area to make the range of systems available for investigation much wider.

All estimations found in the present chapter (and in fact in the whole book) are made up to the order of magnitude. This simplifies calculations and at the same time is sufficient for many practical aims. All calculations are made, for the sake of mathematical simplicity, for Gaussian weight functionals. The correctness of this choice has been discussed in section 4.2 of Chapter 2 (see also a discussion of this question from a different viewpoint in Chapter 10).

5.1 POSITION MONITORING (PATH MEASUREMENT)

In Chapter 4 the measurement amplitude $U_\alpha(t'', q''|t', q')$ was defined as a propagator of the system undergoing a continuous measurement resulting in the output α. This amplitude can be expressed in the form of a weighted (restricted) path integral with the weight functional w_α characterizing the concrete type of measurement and its concrete output. Position monitoring has been considered as an example of a continuous measurement. In this case the measurement output is characterized by a path $\alpha = [a]$ (this is why position monitoring may be called measurement of a path) and the weight functional may be taken to be Gaussian:

$$w_{[a]}[q] = \exp\left(-\frac{2}{T\Delta a^2}\int_0^T \left(q(t) - a(t)\right)^2 dt.\right). \tag{5.1}$$

Here Δa is the error of position measuring, and the measurement interval is taken to be $[0, T]$. The error Δa is taken to be constant. However, the case of time-dependent measurement error can be dealt with quite analogously.

Consider now the measurement amplitude

$$U_{[a]}(q', q) = \langle q'|U_{[a]}|q\rangle = \int_q^{q'} d[q]\, w_{[a]}[q]\, e^{\frac{i}{\hbar}S[q]} \tag{5.2}$$

for position monitoring of a driven harmonic oscillator (i.e. one under the influence of an external force):

$$S[q] = \int_0^T L(q, \dot{q}, t)\, dt, \qquad L = \frac{1}{2}m\dot{q}^2 - \frac{1}{2}m\omega^2 q^2 + q\,F(t).$$

Thus we consider a harmonic oscillator under two types of external influence of different nature: (1) the action of an external force and (2) the back reaction of a measuring device (which monitors the position of this oscillator). The first influence is described by the term $qF(t)$ in the Lagrangian of the oscillator. The second is taken into account with the help of the weight functional $w_{[a]}[q]$ in the integrand of the path integral (5.2).

The importance of this example is seen from the fact that any system near its equilibrium state can be approximated as an oscillator or a system of oscillators. The dynamics of such systems is well investigated, but the influence of continuous measurements on their dynamics is a new and non-trivial problem.

In the above formulas we took the time interval to be $[0, T]$ and the positions of the oscillator before and after continuous measurement to be correspondingly q and q'.

The probabilities of different outputs of a continuous measurement can be found with the help of equations (4.18) and (4.19) of Chapter 4. These formulas contain a final-state density matrix which in turn depends upon an initial state. However, in some cases the dependence on these states is not important. This is when the spread of initial and final states is small so that each of these states may be approximated as being localized at a single point (q and q' correspondingly). The conditions under which this is valid have been investigated in detail in the book by Mensky (1983a).

They are fulfilled for example if the positions of the system at the instants $t = 0, T$ are measured with sufficiently high accuracy.

Without going into the details of these conditions, we shall suggest in the course of the present consideration that the points q, q' are known. Then the probability density for the measurement output $[a]$ is equal to

$$P_{[a]}(q', q) = |U_{[a]}(q', q)|^2. \tag{5.3}$$

To evaluate the measurement amplitude $U_{[a]}$ we can exploit the fact that the effective action arising because of the weight functional (5.1) is quadratic.[1] The result is that the restricted path integral (5.2) coincides with the unrestricted path integral for another harmonic oscillator:

$$U_{[a]}(q', q) = \exp\left(-\frac{2}{T\Delta a^2} \int_0^T a^2 \, dt\right) \int_q^{q'} d[q] \exp\left(\frac{i}{\hbar} \tilde{S}[q]\right) \tag{5.4}$$

where an effective action is introduced:

$$\tilde{S}[q] = \int_0^T \tilde{L}(q, \dot{q}, t) \, dt, \qquad \tilde{L} = \frac{1}{2} m\dot{q}^2 - \frac{1}{2} m\tilde{\omega}^2 q^2 + q \, \tilde{F}(t).$$

It is seen that effective action and effective Lagrangian coincide with those of a harmonic oscillator. The parameters $\tilde{\omega}$, $\tilde{F}(t)$ of this subsidiary (fictitious) oscillator can easily be found and turn out to be complex:

$$\tilde{\omega}^2 = \omega^2 - i\nu^2$$
$$\tilde{F}(t) = F(t) - i\, m\nu^2 a(t)$$

with the imaginary parts determined by the parameter

$$\nu^2 = \frac{4\hbar}{mT\Delta a^2}.$$

It can be seen that the smaller the error Δa of the measurement, the greater is the imaginary part of the effective Lagrangian. This is because a more precise measurement has a stronger influence on a measured system.

The path integral for a driven oscillator can be expressed (see section 3.3 of Chapter 3) in terms of the classical action (the action functional for the classical trajectory). This leads to the following expression for the measurement amplitude:

$$U_{[a]}(q', q) = \left(\frac{m\tilde{\omega}}{2\pi i\hbar \sin \tilde{\omega} T}\right)^{1/2} \exp\left(-2\frac{\langle a^2 \rangle}{\Delta a^2} + \frac{i}{\hbar} \tilde{S}[\tilde{q}_{class}]\right) \tag{5.5}$$

where the 'classical' trajectory of a fictitious complex oscillator

$$m\ddot{\tilde{q}}_{class} + m\tilde{\omega}^2 \tilde{q}_{class} = \tilde{F}, \qquad \tilde{q}_{class}(0) = q, \qquad \tilde{q}_{class}(T) = q'$$

---

[1] The concept of effective action has been defined in section 4.2 of Chapter 4.

and the notation for a time average

$$\langle f \rangle = \frac{1}{T} \int_0^T f(t)\, dt$$

are introduced.

In accordance with the above, we are interested only in the measurement probability (5.3), i.e. in an absolute value of the measurement amplitude. The numerical (normalizing) factor of this amplitude is also unimportant. Thus only the real part of the exponent in (5.5) is needed for the calculation. With the help of integration by parts this real part can be shown to be equal to

$$-2 \frac{\langle |\tilde{q}_{\text{class}} - a|^2 \rangle}{\Delta a^2}.$$

Thus the measurement probability is equal to

$$P_{[a]}(q', q) = J^2 \exp\left(-\frac{4}{\Delta a^2} \langle |\zeta|^2 \rangle\right) \tag{5.6}$$

where the notation

$$\zeta(t) = \tilde{q}_{\text{class}}(t) - a(t)$$

is introduced.

It is easy to show that the function $\zeta(t)$ satisfies the following equation and boundary conditions:

$$m\ddot{\zeta} + m\tilde{\omega}^2 \zeta = \delta F, \qquad \zeta(0) = \delta q, \quad \zeta(T) = \delta q' \tag{5.7}$$

where

$$\delta F = F - F_{[a]}, \qquad \delta q = q - a(0), \quad \delta q' = q' - a(T)$$

with the notation

$$F_{[a]} = m\ddot{a} + m\omega^2 a.$$

A lot of new notation may be discouraging but the final equations (5.6), (5.7) have a simple and natural interpretation. Equation (5.6) shows that probability of the output [a] decreases when the norm

$$\|\zeta\| = \langle |\zeta|^2 \rangle^{1/2}$$

of the subsidiary function $\zeta(t)$ becomes more than the error of measurement, Δa. It can be seen from equation (5.6) that the most probable is the output corresponding to $\zeta(t) \equiv 0$. This in turn corresponds to the case $\delta F \equiv 0$ and $\delta q = \delta q' = 0$. From the definition of δF, δq and $\delta q'$ one can see that the most probable measurement output $[a_0]$ satisfies

$$m\ddot{a}_0 + m\omega^2 a_0 = F, \qquad a_0(0) = q, \quad a_0(T) = q'$$

so that $[a_0] = [q_{\mathrm{class}}]$. This of course could be expected.

One direct conclusion from the result obtained is the possibility of estimating the force acting on the oscillator, knowing the output of the measurement of its path. If the measurement of a path (position monitoring) gives the output $[a]$, then the best estimation for the force is $F_{[a]}$. This is because the functions $F_{[a]}(t)$ and $F(t)$ most probably coincide. More precisely, the greater the difference between $F_{[a]}$ and F, the smaller is the probability of the output $[a]$. We shall give a quantitative formulation of this result below.

The statements of the preceding paragraphs mean that the most probable result of a measurement is the one corresponding to the predictions of classical theory. It is interesting, however, to see how large the deviation of the measurement output from the classical picture may be. This question can be answered if one considers not only the most probable output $[a] = [q_{\mathrm{class}}]$ but also those outputs that have a probability close to the maximum one.[2]

According to equation (5.6), the measurement output $[a]$ is probable enough (its probability is close to maximum) while the norm of the function $\zeta(t)$ is less than or of the order of the measurement error Δa,

$$\|\zeta\| \lesssim \frac{1}{2}\Delta a. \qquad (5.8)$$

The norm $\|\zeta\|$ in turn increases as the values δq, $\delta q'$ and the function δF increase. It is evident that the picture resulting from measurement deviates from the classical prediction when these parameters increase.

The parameters δq, $\delta q'$, δF characterizing the deviation of the measurement output from the classical prediction can be called correspondingly the *deviation of positions* and the *deviation of force*. Recall that q, q' are outputs of measurement of position before and after the continuous measurement. Deflections of the positions δq, $\delta q'$ characterize the deviation of q, q' from the initial and final points $a(0)$, $a(T)$ of the path resulting from the position monitoring (path measurement). When these deviations increase, the probability of the corresponding measurement result $[a]$ decreases. This means that the result of a continuous measurement most probably will be in accordance with the results of instantaneous measurements of position before and after the continuous measurement.

Analogously the function $\delta F(t)$ (deviation of force) characterizes the deviation of the force $F_{[a]}$ (which can be found on the basis of the measurement output) from the actual force F acting on the oscillator during the measurement. The function $\zeta(t)$ depending on q, q' and $F(t)$ may naturally

---

[2] Of course, it would be more correct to talk about probability densities instead of probabilities. Our consideration is not quite rigorous from a mathematical point of view, but it is correct enough for deriving estimations valid up to the order of magnitude, and this is our aim.

be called the *deviation function* and its norm $\|\zeta\|$ the *deviation norm*. They directly characterize the distribution of measurement outputs according to the formula (5.6)

$$P_{[a]}(q', q) = J^2 \exp\left(-\frac{4}{\Delta a^2}\|\zeta\|^2\right). \tag{5.9}$$

5.2 ESTIMATION OF FORCE ACTING ON AN OSCILLATOR

We saw in section 5.1 that the probability of the output $[a]$ of the measurement of a path (position monitoring) depends on two numbers δq, $\delta q'$ (position deviations) and a function $\delta F(t)$ (a force deviation). All these parameters characterize the deviation from classical predictions of the picture resulting from the measurement output. Let us now accept

$$\delta q = \delta q' = 0 \tag{5.10}$$

and investigate the dependence of the output probability $P_{[a]}$ on the force deviation δF. This will allow us to find the precision of estimation of a force on the basis of the measurement result.

The formulation of a typical practical problem is as follows: given the output $[a]$ of position monitoring, what is possible deviation of the best estimation $F_{[a]}$ of a force from the actual force F acting on the oscillator?

To answer this question, we can suppose that actual force is F, find the force deviation $\delta F = F - F_{[a]}$, then find deviation function $\zeta(t)$ from

$$m\ddot{\zeta} + m\tilde{\omega}^2\zeta = \delta F, \qquad \zeta(0) = \zeta(T) = 0 \tag{5.11}$$

and determine the probability density $P_{[a]}$ by the formula (5.9).

If the resulting value $P_{[a]}$ is close to maximum, then the actual force could be equal to $F_{[a]}$ with large probability. If $P_{[a]}$ turns out to be much less, then the force F cannot be equal to $F_{[a]}$ (the probability of this event is small).[3] In practice the inequality (5.8) is sufficient for estimating those forces F which are probable in the light of information about the measurement output $[a]$.

To obtain a clearer notion about the precision of the force estimation, let us fix the form of the force deviation $\delta F(t)$. Suppose for example that this function is a harmonic one. The null boundary conditions (5.10) then fix the frequency and phase of the harmonic function, so that

$$\delta F(t) = \delta F_\Omega \sin \Omega t \tag{5.12}$$

---

[3] For a stricter evaluation of the probability of the force F on the basis of the measurement result $[a]$ a sort of Bayesian process is needed (see Helstrom 1976). However, the above consideration is enough for a rough estimation.

with $\Omega = n\pi/T$ (integer n). If n is fixed, the only arbitrariness in the function δF is its amplitude δF_Ω. Our task is to estimate the maximum value of $|\delta F_\Omega|$ for which the inequality (5.8) is still valid so that $F = F_{[a]} + \delta F$ is probable in the light of information that the measurement result is $[a]$.

The equation (5.11) can be easily solved with the right-hand-side (5.12) to give

$$\zeta(t) = \frac{\delta F_\Omega}{m(\tilde{\omega}^2 - \Omega^2)} \sin \Omega t, \qquad \|\zeta\|^2 = \frac{\delta F_\Omega^2}{2m^2[(\Omega^2 - \omega^2)^2 + \nu^4]}.$$

Substituting the derived expression for a deviation norm in (5.8), one obtains a restriction for δF_Ω. According to this inequality the maximum value of δF_Ω is defined by

$$\delta F_{\Omega\,\text{max}}^2 = \frac{1}{2} m^2 (\Omega^2 - \omega^2)^2 \left(\Delta a^2 + \frac{\Delta a_\Omega^4}{\Delta a^2} \right) \qquad (5.13)$$

where

$$\Delta a_\Omega^2 = \frac{4\hbar}{mT|\Omega^2 - \omega^2|}.$$

Our final conclusion is that the force deviation of the harmonic form (5.12) is probable only for

$$|\delta F_\Omega| \lesssim \delta F_{\Omega\,\text{max}} \qquad (5.14)$$

with the maximum value determined by equation (5.13). This means that (5.13) is the error of estimation of a force (at the given frequency) on the basis of the continuous measurement output. Of course, a continuous measurement of the given type (position monitoring, or path measurement) is meant.

We proved the inequality (5.14) for the frequency equal to $\Omega = n\pi/T$ and for special choice of the phase of the harmonic function $\delta F(t)$. However, the same inequality can be proved to be valid for any frequency and phase, excluding frequencies differing by less than a value of the order of π/T from the resonance frequency ω. Thus if

$$|\Omega - \omega| \gtrsim \frac{\pi}{T}$$

and

$$\delta F(t) = \delta F_\Omega \sin(\Omega t + \Phi) \qquad (5.15)$$

the inequality (5.14) is valid for all measurement outputs having reasonable probability.

Let us formulate once more the results obtained above.

If the path measurement of the oscillator (performed with the error Δa) gives the output $[a]$, then the best estimation of the force F acting on the oscillator is

$$F_{[a]}(t) = m\ddot{a} + m\omega^2 a.$$

However, an actual force $F(t)$ may differ from this estimation. The error of the estimation at frequency Ω (differing from the resonance frequency ω by more than π/T) cannot be more than $\delta F_{\Omega\,\text{max}}$. This means that if the deviation $\delta F(t)$ of an actual force F from its estimation $F_{[a]}$ is of the form (5.15), then its amplitude δF_Ω satisfies the inequality (5.14). The case of frequencies close to the resonance frequency requires a more accurate analysis including non-zero position deviations δq, $\delta q'$. This will be done later on (see section 5.3).

The formula (5.13) shows that there are two different regimes of measurement (with respect to the given frequency Ω), $\Delta a \gg \Delta a_\Omega$ and $\Delta a \ll \Delta a_\Omega$, and also the special regime at the boundary between them, $\Delta a \simeq \Delta a_\Omega$. Let us consider them separately.

If the measurement is rough, $\Delta a \gg \Delta a_\Omega$, then the second term on the right-hand side of the formula (5.13) is negligible, giving

$$\delta F_{\Omega\,\text{max}} \simeq \frac{1}{\sqrt{2}} m |\Omega^2 - \omega^2| \Delta a. \qquad (5.16)$$

This formula is natural in the sense that the error of the force estimation decreases with decreasing measurement error. Moreover, the error (5.16) can be easily shown to correspond to classical theory.

Indeed, in classical theory there is one-to-one correspondence (for the fixed boundary conditions) between the force $F(t)$ and the path $q(t)$. A difference in paths of the form $\Delta q(t)$ leads to a difference in forces equal to

$$\Delta F(t) = m(\Delta\ddot{q} + \omega^2 \Delta q).$$

The difference of the harmonic form of the frequency Ω arises from the harmonic difference of paths of the same frequency, and amplitudes of the corresponding harmonic functions can be connected as follows:

$$\Delta F_\Omega = m |\Omega^2 - \omega^2| \Delta q_\Omega.$$

If a path is measured with the error Δa then Δq_Ω is restricted by a value of the order of Δa, and consequently ΔF_Ω is restricted by a value of the order of (5.16).

Thus the measurement of a path performed with the error $\Delta a \gg \Delta a_\Omega$ has a purely classical character at the frequency Ω. This can be called the *classical regime of measurement*.

The opposite case of a sufficiently precise measurement, $\Delta a \ll \Delta a_\Omega$, leads to the first term on the right-hand side of equation (5.13) being negligible. This gives

$$\delta F_{\Omega\,\text{max}} \simeq \frac{2\sqrt{2}\hbar}{T\Delta a}. \tag{5.17}$$

The *quantum regime of measurement* that arises is paradoxical in the sense that an increasing precision of the measurement leads in this regime to a less precise estimation of the force.

The error in estimation in this case is a consequence of quantum effects (quantum fluctuations, or *quantum measurement noise*), i.e. of an unavoidable back reaction of a measuring device on the measured system. In complete correspondence with the principles of quantum mechanics this back reaction is greater for more precise measurement. Moreover, the formula (5.17) is in direct correspondence with the Heisenberg uncertainty principle. This can be seen if one denotes the product $T\delta F_\Omega$ as indeterminacy in some effective momentum (for a given frequency), Δp_Ω. Then it is seen from equation (5.17) that product $\Delta a \Delta p_\Omega$ is of the order of $\hbar$.

Consider now the complete range of variation of the measurement error Δa. When Δa is large, the classical regime of measurement occurs. Then diminishing Δa leads (in complete correspondence with the classical intuition) to a decrease in the force estimation error δF_Ω. However, this is only until Δa becomes less than the threshold value Δa_Ω. When this threshold is crossed, the quantum regime is entered. A further decrease in Δa leads paradoxically to an increase in the force estimation error δF_Ω as a consequence of quantum fluctuations, or of the quantum measurement noise.

The intermediate regime between the classical and quantum ones, $\Delta a \simeq \Delta a_\Omega$, allows one to achieve minimum error in the force estimation. This optimal regime arises when the first (classical) and the second (quantum) terms on the right-hand side of equation (5.13) are equal to each other. The resulting optimal (minimum) value of the force estimation error is equal to

$$\delta F_\Omega^{\text{opt}} \simeq m|\Omega^2 - \omega^2|\Delta a_\Omega = 2\sqrt{\frac{m\hbar|\Omega^2 - \omega^2|}{T}}. \tag{5.18}$$

This formula is valid for a frequency Ω sufficiently far from the resonance frequency ω (greater than a distance of the order of π/T). The case of resonance measurement will be considered in section 5.3.3.

5.3 SPECTRAL MEASUREMENTS OF AN OSCILLATOR

The model of measurement considered in the preceding sections is not realistic in one respect: effective monitoring at all frequencies is impossible. Therefore it is necessary to consider measurements performed in a restricted frequency interval. On the other hand, an important class of measurements is measuring of definite frequency components of a system path. For example, measurement of an oscillator is often performed in a narrow frequency interval near the resonance frequency. A measurement of this type will be investigated in this section for an harmonic oscillator as the measured system.

5.3.1 Formulation of the Problem

Let us expand the function $q(t)$ representing a path of the oscillator in a Fourier series at the interval $[0, T]$:

$$q(t) = \sum_{n=1}^{\infty} q_n \sin \Omega_n t \qquad (5.19)$$

where

$$\Omega_n = \frac{n\pi}{T}$$

and consider the measurement of components q_n with the errors Δa_n as a continuous measurement. Let us call this type of measurement the spectral measurement.

An advantage of the approach under consideration is that no concrete model of a measurement device is required. However, it is useful to get some idea of the physical conditions under which the considered measurement is feasible. Let us remark in this connection that any measuring device monitoring position with the error Δa but reacting (like any real device) only at frequencies in a restricted interval performs in fact a sort of spectral measurement.

If $I = [\Omega_1, \Omega_2]$ is the interval of frequencies in which the device is effective, then this measurement may be considered to be a spectral measurement of the frequency components q_n, $\Omega_n \in I$, with equal errors $\Delta a_n = \Delta a$ while the frequency components $q_{n'}$, $\Omega_{n'} \notin I$ outside this interval are measured with the error $\Delta a_{n'} = \infty$. A more realistic representation of monitoring by a real device is the choice of Δa_n close to Δa for frequencies in the middle of the interval I, increasing near the edges of the interval I and very large outside this interval.

The simplest case is measurement of a single frequency component q_n with the error Δa_n. This measurement can be described by a weight functional of the form

$$w_{a_n, \Delta a_n}[q] = \exp\left(-\frac{(q_n - a_n)^2}{\Delta a_n^2}\right). \qquad (5.20)$$

In a general case frequency components q_n, $n = 1, 2, \ldots$, are measured with the corresponding errors Δa_n. Such a measurement can be described by the functional

$$w_{a_1 a_2 \ldots}[q] = \prod_{n=1}^{\infty} \exp\left(-\frac{(q_n - a_n)^2}{\Delta a_n^2}\right) = \exp\left(-\sum_{n=1}^{\infty} \frac{(q_n - a_n)^2}{\Delta a_n^2}\right). \quad (5.21)$$

The frequency components q_n can be expressed as integrals of a function $q(t)$:

$$q_n = \frac{2}{T} \int_0^T q(t) \sin \Omega_n t \, dt. \quad (5.22)$$

This gives an explicit expression of the functional $w_{a_1 a_2 \ldots}[q]$ in terms of the function $q(t)$.

For the transition from series to integrals the *Parseval formula*

$$(f, g) = \frac{1}{T} \int_0^T f^*(t) g(t) \, dt = \sum_{n=1}^{\infty} f_n^* g_n \quad (5.23)$$

may often be of help. If the errors of measurement of all frequency components coincide, $\Delta a_1 = \Delta a_2 = \ldots = \Delta a$, then the functional (5.21) can be shown (with the help of (5.23)) to coincide with the functional (5.1). Thus position monitoring is a special case of a spectral measurement. This is why some conclusions of the present section will allow us to make the results of section 5.2 more precise.

5.3.2 *Results of Evaluation*

The path integrals

$$U_{a_n, \Delta a_n}(q', q) = \int_q^{q'} d[q] \, w_{a_n, \Delta a_n}[q] \, e^{\frac{i}{\hbar} S[q]} \quad (5.24)$$

$$U_{a_1, a_2, \ldots}(q', q) = \int_q^{q'} d[q] \, w_{a_1, a_2, \ldots}[q] \, e^{\frac{i}{\hbar} S[q]} \quad (5.25)$$

with the weight functionals (5.20), (5.21) can be evaluated with the help of the spectral representation of a path integral (Feynman and Hibbs 1965, see also section 3.3 of Chapter 3). This means that the measure $d[q]$ is represented in the form

$$d[q] = \prod_{n=1}^{\infty} dq_n$$

so that a path integral is reduced to an infinite-dimensional integral over frequency components.

All integrals arising in the calculation with the functionals (5.20), (5.21) are of Gaussian type and can be evaluated precisely. The resulting amplitude $U_{a_1 a_2 \ldots}$ allows one to find probability densities

$$P_{a_n, \Delta a_n} = |U_{a_n, \Delta a_n}|^2$$
$$P_{a_1 a_2 \ldots} = |U_{a_1 a_2 \ldots}|^2.$$

It can be shown with the help of a straightforward though cumbersome calculation that in the simple case of measurement of a single frequency component (5.20)

$$P_{a_n, \Delta a_n} = \exp\left(-2\frac{(a_n - a_n^{\text{class}})^2}{\Delta a_n^2 + \frac{(\Delta a_n^{\text{opt}})^4}{\Delta a_n^2}}\right) \tag{5.26}$$

and in a general case

$$P_{a_1 a_2 \ldots} = \prod_{n=1}^{\infty} P_{a_n, \Delta a_n} = \exp\left(-2\sum_{n=1}^{\infty} \frac{(a_n - a_n^{\text{class}})^2}{\Delta a_n^2 + \frac{(\Delta a_n^{\text{opt}})^4}{\Delta a_n^2}}\right). \tag{5.27}$$

In these formulas the following entity emerged:

$$a_n^{\text{class}} = (\Omega_n^2 - \omega_n^2)^{-1}\left(-\frac{F_n}{m} + \frac{2\Omega_n}{T}\left(q - (-1)^n q'\right)\right).$$

This coincides in fact with the measurement result predicted by classical theory (depending of course on the boundary conditions q, q' and the corresponding frequency components F_n of the force $F(t)$). In addition, we used the notation

$$(\Delta a_n^{\text{opt}})^2 = \frac{4\hbar}{mT|\Omega_n^2 - \omega^2|}.$$

Let us consider what conclusions follow from the probability distributions for measurement outputs. We shall begin with the simplest case of measuring a single frequency component q_n. Then the probabilities of different outputs of measurement are described by the formula (5.26). The most probable measurement output a_n is a_n^{class}. However, a_n can differ from a_n^{class} by the quantity

$$\delta a_n = \sqrt{\Delta a_n^2 + \frac{(\Delta a_n^{\text{opt}})^4}{\Delta a_n^2}}. \tag{5.28}$$

In the classical regime of measurement, $\Delta a_n \gg \Delta a_n^{\text{opt}}$, one has

$$\delta a_n = \Delta a_n,$$

just as in classical measurement theory. In the quantum regime, $\Delta a_n \ll \Delta a_n^{\mathrm{opt}}$, one obtains

$$\delta a_n = \frac{(\Delta a_n^{\mathrm{opt}})^2}{\Delta a_n}.$$

The variance of outputs then increases as the measurement becomes more accurate. This paradoxical feature is a consequence of the unavoidable back reaction of the measuring apparatus on a quantum system. The optimal regime corresponds to the choice $\Delta a_n \simeq \Delta a_n^{\mathrm{opt}}$. In this case

$$\delta a_n^{\mathrm{opt}} = \left(\frac{8\hbar}{mT|\Omega_n^2 - \omega^2|} \right)^{1/2}. \tag{5.29}$$

5.3.3 *Estimation of the Acting Force*

Measurement of the coordinate frequency component q_n allows one to estimate the frequency component F_n of the force $F(t)$ acting on the oscillator during the measurement. This estimation F_n^{est} can be derived from the distribution (5.26) if one takes into account that this distribution reaches a maximum when $a_n = a_n^{\mathrm{class}}$. Having obtained the output a_n, the experimentalist should estimate the external force from the formula

$$F_n^{\mathrm{est}} = m(\omega^2 - \Omega_n^2)a_n + \frac{2m\Omega_n}{T}(q - (-1)^n q'). \tag{5.30}$$

However, according to the distribution (5.26), a_n can differ from a_n^{class} by the quantity (5.28). Therefore, the error of estimation of the force will be

$$\delta F_n = m|\Omega_n^2 - \omega^2|\delta a_n \tag{5.31}$$

in a general case and

$$\delta F_n^{\mathrm{opt}} = \left(\frac{8\hbar m|\Omega_n^2 - \omega^2|}{T} \right)^{1/2} \tag{5.32}$$

in the optimal regime.

In the case of the resonance measurement, $\Omega_n \to \omega$, the boundary conditions q, q' cannot be considered to be known precisely, since the quantum measurement noise depends essentially on the error Δq with which they are known. Taking into account indeterminacy in the second term in (5.30) gives, instead of (5.32), the following formula:

$$\delta F_n^{\mathrm{opt}} = \left(\frac{8\hbar m|\Omega_n^2 - \omega^2|}{T} \right)^{1/2} + \frac{4m\Omega_n}{T}\Delta q. \tag{5.33}$$

If the oscillator is in the coherent state before and after the measurement, then

$$\Delta q = \Delta q_{\text{coher}} = (\hbar/m\omega)^{1/2}.$$

Let the period of measurement T be chosen close to $n\pi/\omega$ so that Ω_n becomes close to ω and the first term in equation (5.33) becomes small while the second term dominates. Then this formula gives the restriction for resonance measurement in the case when the initial state is a coherent one:

$$\delta F_{\text{res}}^{\text{opt}} = 4\frac{\sqrt{\hbar m\omega}}{T}. \tag{5.34}$$

This is the so-called *standard quantum limit*.

We see that the natural indeterminacy of coordinates characteristic of the coherent (the most common) state does not allow $\delta F_n^{\text{opt}} = 0$ to be achieved as might seem possible from equation (5.32). This restriction can, however, be overcome with the help of very precise measurement of positions before and after the continuous measurement.

Let one measure the coordinates q, q' at the instants $t = 0, T$ with high accuracy, so that the indeterminacy of coordinates at these instants becomes less than Δq_{coher}. Then the second term in (5.33) can be made arbitrarily small. If, in addition, the time interval T is chosen in such a way that Ω_n differs from ω by a still smaller amount, then the standard quantum limit can be overcome. In this case one has in fact a *stroboscopic regime of measurement* as proposed by Braginsky et al (1978).

Consider now the case when several frequency components q_n are measured simultaneously (as described by (5.25)). In this case the measurement output allows one to estimate the corresponding components of the force, F_n^{est}. This in turn allows one to restore the force $F^{\text{est}}(t)$ as a function of time, with the help of a formula of the type of (5.19). Then the mean square deviation of the actual force from its estimation

$$\langle (F - F^{\text{est}})^2 \rangle = \frac{1}{T} \int_0^T dt \left(F(t) - F^{\text{est}}(t) \right)^2$$

can be evaluated with the help of the Parseval formula (5.23). If the considered frequency interval is not very wide (compared with π/T) it can be shown that the mean square deviation does not exceed the value

$$\langle (F - F^{\text{est}})^2 \rangle = \frac{4\hbar m}{T} |\Omega^2 - \omega^2|. \tag{5.35}$$

For the frequency band that is symmetric around the resonance frequency ω one has

$$\langle (F - F^{\text{est}})^2 \rangle = \frac{4\hbar m\omega\Delta\Omega}{T}. \tag{5.36}$$

Let us apply the latter formula to the case of detecting gravitational waves with the help of a Weber-type antenna. The Weber bar (or rather one mode of its oscillations) can be represented in this case as a harmonic oscillator. The effective force acting on this oscillator is

$$F = -mc^2\, R^1{}_{010}\, q.$$

The component of the curvature tensor arising here can be expressed through the dimensionless variation of the metric, h, as follows:

$$c^2\, R^1{}_{010} = -\frac{1}{2}\ddot{h}.$$

The second derivative of the metric h is of the order of $\Omega^2 h$.
   This gives, to the order of magnitude,

$$F \simeq \frac{1}{2}ml\Omega^2 h$$

where l is the length of the Weber bar.
   The formula (5.35) gives then for a minimum of detectable variation of the metric

$$\delta h^2 = \frac{16\hbar|\Omega^2 - \omega^2|}{mTl^2\Omega^4}. \tag{5.37}$$

If it is not necessary to measure a wide frequency band, then one can put $\Delta\Omega$ to be the minimum possible, π/T. Then, for $\Omega \simeq \omega$,

$$\delta h = \frac{4}{Tl\omega}\sqrt{\frac{\pi\hbar}{m\omega}}. \tag{5.38}$$

For $m = 10^6$ g, $\omega = 10^4\,\mathrm{s}^{-1}$, $l = 10^2$ cm and $\omega T \simeq 2\pi$, one has $\delta h \simeq 10^{-21}$.
   This value gives an absolute limit on the sensitivity of a gravitational-wave antenna of Weber type under the condition that the measurement of its coordinate is used to estimate the gravitational-wave signal. There are ways to overcome this standard quantum limit. One of them is the stroboscopic measurement mentioned above. The other ways are based on quantum-nondemolition measurements (see Chapter 6).

5.3.4 System of Connected Oscillators

The system of connected harmonic oscillators is a universal model for any oscillating system in a state close to its equilibrium state. For example, the Weber bar is usually connected to an electric LC circuit, the measurement of which gives indirect information about the behaviour of the bar and therefore about gravitational waves. All of two or more connected

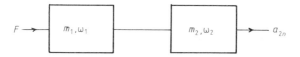

Figure 5.1: Two oscillators in interaction. The force F acts on the first of them; the nth frequency component of the path of the second is measured.

oscillators may display quantum properties. Therefore the problem of measurements in a system of connected quantum oscillators naturally arises. We shall consider here the case of two oscillators (see Mensky 1983a), suggesting that one of them is under action of an external force and the other undergoes the continuous measurement (see figure 5.1).

The measurement amplitude is expressed in our case in the form of a path integral in the coordinates of the two oscillators:

$$U_{a_{2n}, \Delta a_{2n}}(q', q) = \int_{q_1}^{q_1'} d[q_1] \int_{q_2}^{q_2'} d[q_2] \, w_{a_{2n}, \Delta a_{2n}}[q_2] \, e^{\frac{i}{\hbar} S[q_1, q_2]}$$

where we have supposed that only the coordinate of the second oscillator is measured. The corresponding Lagrangian has the form

$$L = \frac{1}{2} m_1 (\dot{q}_1^2 - \omega_1^2 q_1^2) + \frac{1}{2} m_2 (\dot{q}_2^2 - \omega_2^2 q_2^2) - g q_1 q_2 + F q_1. \qquad (5.39)$$

The weight factor describing the measurement of the nth frequency component of the coordinate of the second oscillator has the form

$$w_{a_{2n}, \Delta a_{2n}}[q] = \exp\left(-\frac{(q_{2n} - a_{2n})^2}{\Delta a_{2n}^2}\right).$$

Now the path integral can be evaluated in just the same way as for a single oscillator. The result of this calculation is

$$P_{a_{2n}, \Delta a_{2n}} = \exp\left(-2 \frac{(a_{2n} - a_{2n}^{\text{class}})^2}{\delta a_{2n}^2}\right) \qquad (5.40)$$

with the notation

$$\delta a_{2n}^2 = \Delta a_{2n}^2 + \frac{(\Delta a_{2n}^{\text{opt}})^4}{\Delta a_{2n}^2}$$

$$(\Delta a_{2n}^{\text{opt}})^2 = \frac{4\hbar m_1 |\Omega_n^2 - \omega_1^2|}{T |m_1 m_2 (\Omega_n^2 - \omega_1^2)(\Omega_n^2 - \omega_2^2) - g^2|}. \qquad (5.41)$$

The quantity a_{2n}^{class} here corresponds to the classical solution of the problem. Analysis of the distribution obtained is obviously analogous to that performed above for a single oscillator. This analysis shows that the best estimation of the force acting on the system of oscillators coincides with the classical estimation, and the error of the estimation is equal to

$$\delta F_n^{\text{opt}} = \frac{\hbar m_1}{T} \left| \frac{(\Omega_n^2 - \omega_1^2)(\Omega_n^2 - \nu_1^2)(\Omega_n^2 - \nu_2^2)}{(\omega_1^2 - \nu_1^2)(\omega_2^2 - \nu_2^2)} \right| \qquad (5.42)$$

where the resonance frequencies of the system of two oscillators are introduced:

$$\nu_{1,2}^2 = \frac{1}{2}(\omega_1^2 + \omega_2^2) \mp \sqrt{\frac{1}{4}(\omega_2^2 - \omega_1^2)^2 + \frac{g^2}{m_1 m_2}}.$$

If the frequency Ω_n approaches one of the characteristic frequencies ω_1, ν_1, ν_2, then the formula (5.42) becomes invalid, since uncertainty of the boundary conditions for q_1, q_2 must be taken into account in this case. If the system is in a coherent state before and after the measurement, then one has for each of these characteristic frequencies

$$\delta F^{\text{opt}}(\omega_1) = \frac{1}{T}\sqrt{2\pi\hbar m_1 \omega_1}$$

$$\delta F^{\text{opt}}(\nu_1) = \frac{1}{T}\sqrt{2\pi\hbar m_1 \nu_1} \times \sqrt{\frac{\nu_2^2 - \nu_1^2}{\omega_2^2 - \nu_1^2}}$$

$$\delta F^{\text{opt}}(\nu_2) = \frac{1}{T}\sqrt{2\pi\hbar m_1 \nu_2} \times \sqrt{\frac{\nu_2^2 - \nu_1^2}{\omega_1^2 - \nu_1^2}}.$$

The system of many interacting oscillators can be considered quite analogously.

5.4 EVOLUTION OF A HARMONIC OSCILLATOR SUBJECT TO SPECTRAL MEASUREMENT

The evolution law of a quantum system subject to continuous measurement was formulated in sections 4.3 and 4.4 of Chapter 4. Taking the starting instant to be $t = 0$ and final instant $t = T$, we have for this law

$$\rho_T = \int d\mu(\alpha)\, U_\alpha\, \rho_0 U_\alpha^\dagger \qquad (5.43)$$

or, in the coordinate representation of the density matrix,

$$\rho_T(q_1', q_2') = \int dq_1 \int dq_2\, U(q_1', q_1 | q_2', q_2)\, \rho_0(q_1, q_2) \qquad (5.44)$$

where the following notation is introduced

$$U(q_1', q_1 | q_2', q_2) = \int_{q_1}^{q_1'} d[q_1] \int_{q_2}^{q_2'} d[q_2]$$

$$\times w([q_1], [q_2]) \exp\left(\frac{i}{\hbar}(S[q_1] - S[q_2])\right) \quad (5.45)$$

$$w([q_1], [q_2]) = \int d\mu(\alpha) \, w_\alpha[q_1] w_\alpha[q_2]. \quad (5.46)$$

The measure $d\mu(\alpha)$ has to be chosen in such a way that the generalized unitarity condition be fulfilled:

$$\int dq_1' \, U(q_1', q_1 | q_1', q_2) = \delta(q_1 - q_2). \quad (5.47)$$

Consider now a special case of spectral measurement described in section 5.3. In this case the measurement output α is expressed by the set $a_1, a_2, \ldots$ of numbers (values of frequency components of a path). The measure $d\mu(\alpha)$ may be evidently reduced to integration (with some weight) over the variables $a_1, a_2, \ldots$. The calculation to be performed in the present section will show that the weight should be taken equal to unity. Thus we shall choose

$$d\mu(\alpha) = d\mu(a_1, a_2, \ldots) = da_1 da_2 \ldots \quad (5.48)$$

to verify afterwards that this choice is appropriate. With this choice equation (5.46) reads

$$w([q_1], [q_2]) = \int da_1 \int da_2 \ldots w_{a_1, a_2, \ldots}[q_1] w_{a_1, a_2, \ldots}[q_2]. \quad (5.49)$$

The weight functional for the spectral measurement is (5.21)

$$w_{a_1 a_2 \ldots}[q] = \exp\left(-\sum_{n=1}^{\infty} \frac{(q_n - a_n)^2}{\Delta a_n^2}\right). \quad (5.50)$$

Substituting this expression for the weight functional in (5.49) and performing Gaussian integration, one has

$$w([q_1], [q_2]) \sim \exp\left(-\frac{1}{2}\sum_{n=1}^{\infty} \frac{(q_{1n} - q_{2n})^2}{\Delta a_n^2}\right).$$

Here and in what follows we omit the numerical factor, which will be recovered at the end of the calculation with the help of the normalization condition.

To calculate the integrals over $[q_1]$ and $[q_2]$ in equation (5.45), it is convenient to change to the variables z_1 and z_2 shifted by the classical trajectories with the corresponding boundary conditions:

$$q_1 = z_1 + \eta_1, \qquad \ddot{\eta}_1 + \omega^2 \eta_1 = 0, \quad \eta_1(0) = q_1, \quad \eta_1(T) = q_1',$$
$$q_2 = z_2 + \eta_2, \qquad \ddot{\eta}_2 + \omega^2 \eta_2 = 0, \quad \eta_2(0) = q_2, \quad \eta_2(T) = q_2'.$$

Then the expression (5.45) takes the form

$$\exp\left(\frac{i}{\hbar}\left(S[\eta_1] - S[\eta_2]\right)\right) \int_0^0 d[z_1] \int_0^0 d[z_2]$$
$$\times \exp\left(\frac{i}{\hbar}\left(S[z_1] - S[z_2]\right) - \frac{1}{2}\sum_{n=1}^{\infty} \frac{(z_{1n} + \eta_{1n} - z_{2n} - \eta_{2n})^2}{\Delta a_n^2}\right).$$

We have here integrals over paths $[z_1]$, $[z_2]$ with null boundary conditions at the instants $t = 0, T$. Such integrals can be evaluated with the help of the spectral representation of path integrals (Feynman and Hibbs 1965, see also section 3.3 of Chapter 3), i.e. by going from integration over paths $[z_i]$ to integration over the spectral components of these paths, $z_{in}, n = 1, 2, \ldots$. As a result we arrive at integrals of Gaussian type, and for the function U this gives the expression

$$\exp\left(\frac{i}{\hbar}\left(S[\eta_1] - S[\eta_2]\right) - \frac{1}{2}\sum_{n=1}^{\infty}\frac{(\eta_{1n} - \eta_{2n})^2}{\Delta a_n^2}\right).$$

Using an explicit form of the classical trajectory, we can readily obtain for the function U the final expression

$$U(q_1', q_1 | q_2', q_2)$$
$$= N \exp\left(\frac{i}{\hbar}\left(\Phi(q_1, q_1') - \Phi(q_2, q_2')\right) + \Lambda(q_1 - q_2, q_1' - q_2')\right) \quad (5.51)$$

where

$$\Phi(q, q') = \frac{m\omega \cos\omega T}{2\sin\omega T}(q^2 + q'^2) - \frac{m\omega}{\sin\omega T}qq' + q'F_\omega + qG_\omega$$
$$\Lambda(u, v) = -\kappa(u^2 + v^2) + 2\sigma uv$$

with the following notation:

$$F_\omega = \frac{1}{\sin\omega T}\int_0^T dt\, f(t)\sin\omega t, \qquad G_\omega = \frac{1}{\sin\omega T}\int_0^T dt\, f(t)\sin\omega(T - t),$$

$$\kappa = \frac{2}{T^2}\sum_{n=1}^{\infty}\frac{\Omega_n^2}{\Delta a_n^2(\Omega_n^2 - \omega^2)^2}, \qquad \sigma = \frac{2}{T^2}\sum_{n=1}^{\infty}\frac{(-1)^n\Omega_n^2}{\Delta a_n^2(\Omega_n^2 - \omega^2)^2}.$$

The normalization constant

$$N = m\omega/2\pi\hbar \sin\omega T$$

can be found from the requirement that the function (5.51) should satisfy the generalized unitarity condition (5.47). Note that this condition could be satisfied by no choice of the constant N had we chosen the measure $d\mu(\alpha)$ on the space of results of the measurement incorrectly. Therefore, the fulfillment of the generalized unitarity condition means that the choice (5.48) adopted in our calculation was correct.

Equation (5.44) with the superpropagator defined by equation (5.51) describes evolution of the quantum harmonic oscillator undergoing continuous measurement of the type of spectral measurement. A specific feature of such an evolution is that a pure state of the system is converted after the evolution into a mixed state. This is a consequence of measurement, leading to loss of quantum coherence. A simple example of this phenomenon has been considered in section 4.4 of Chapter 4. The special case $\omega = 0$ of the formula (5.44) was applied there to describe a free particle undergoing continuous measurement. Instead of the general case of a spectral measurement, a special case was taken there corresponding to $\Delta a_1 = \Delta a_2 = \ldots = \Delta a$. We saw in section 5.3 that such a measurement coincides in fact with position monitoring, or path measurement.

5.5 MEASUREMENT OF A PATH OF A NONLINEAR OSCIL-LATOR

In the preceding sections we have considered continuous measurements of harmonic oscillators. As has been mentioned, in many situations real systems may actually be treated as harmonic oscillators. However, in many cases this approximation is insufficient and the nonlinear properties of real systems should be taken into account. This is important for example when investigating quantum systems whose classical prototypes possess chaotic properties.[4] Here we shall consider the simplest example of a nonlinear system: the nonlinear oscillator.[5]

We shall use a technique based upon numerical integration of an effective Schrödinger equation and apply it to continuous monitoring of the position (measurement of a path) of a nonlinear oscillator.

---

[4] Several numerical experiments have shown that quantum–classical correspondence in chaotic dynamics seems to be anomalous (Casati et al 1979, Chirikov et al 1981). However, investigation of this phenomenon may require a quantum measurement process (Shepelyansky 1983). Adachi et al (1989) investigated an example of a chaotic system in a quantum regime under instantaneous measurements.

[5] The results to be presented here were obtained in collaboration with Roberto Onofrio and Carlo Presilla, see Mensky et al (1991).

If the system moves between the points of the space-time $(q, 0)$ and (q', T) the result of evolution $U_{[a]}(T, q'|0, q) = U_{[a]}(q', q)$ depends on the measurement output $[a] = \{a(t) \mid 0 \le t \le T\}$. This can be interpreted in two different ways. Firstly, it represents the probability amplitude for the measurement output $[a]$ given the positions of the system q and q' before and after the measurement. Secondly, it can be understood as a propagator for the system subject to the continuous measurement given the output of the measurement. The restriction of the path integral to the set of paths compatible with the measurement output can be effectively done with the help of a weight functional $w_{[a]}[q]$ depending on the measurement output $[a]$ and decaying outside the set of paths compatible with $[a]$:

$$U_{[a]}(q', q) = \int d[q] \exp\left(\frac{i}{\hbar} \int_0^T L(q, \dot{q}, t)\, dt\right) w_{[a]}[q].$$

For example, if the coordinate q is monitored with the error Δa and the result of the measurement is $[a]$, the weight functional $w_{[a]}[q]$ selects the paths $q(t)$ in the corridor centred around $[a]$ and having a width Δa. A simple choice is a Gaussian weight functional

$$w_{[a]}[q] = \exp\left(-\frac{2}{T\Delta a^2} \int_0^T (q(t) - a(t))^2\, dt\right).$$

The restricted path integral is then

$$U_{[a]}(q', q) = \int d[q] \exp\left(\frac{i}{\hbar} \int_0^T L(q, \dot{q}, t)\, dt - \frac{2}{T\Delta a^2} \int_0^T (q - a)^2\, dt\right).$$

It is important (see section 4.2 of Chapter 4) that the resulting path integral may be considered as describing a free (i.e. not measured) system but with an effective Lagrangian having an imaginary term due to the measurement:

$$L_{\text{eff}}(q, \dot{q}, t) = L(q, \dot{q}, t) + \frac{2i\hbar}{T\Delta a^2} (q - a(t))^2. \tag{5.52}$$

The imaginary term produces a decrease in the density of the system in the configuration space far from $q(t) = a(t)$. Finally, the effective Lagrangian is time-dependent even if the original Lagrangian is not.

In order to estimate the probability distribution for the measurement outputs of a physical system we consider the convolution

$$I_{[a]} = \langle \psi_1 | U_{[a]} | \psi_0 \rangle = \int \int \psi_1^*(q') U_{[a]}(q', q) \psi_0(q)\, dq\, dq'.$$

According to the first interpretation of $U_{[a]}$ the quantity $I_{[a]}$ is a probability amplitude for the measurement to give the output $[a]$ under the condition that the system has been in the state ψ_0 before the measurement and in the state ψ_1 after a time T. The probability distribution for the measurement output is then

$$P_{[a]} = \frac{|I_{[a]}|^2}{\int d[a] |I_{[a]}|^2}.$$

Note that $P_{[a]}$ depends upon the instrumental uncertainty Δa.

The amplitude $I_{[a]}$ can be also written as a scalar product

$$I_{[a]} = \langle \psi_1 | \Psi_{[a]}(T) \rangle$$

where

$$\Psi_{[a]}(q', T) = \int U_{[a]}(q', q) \psi_0(q) \, dq.$$

According to the second interpretation of $U_{[a]}$ the wavefunction $\Psi_{[a]}(q, T)$ represents the evolution at time T of the state $\psi_0(q)$ under the action of a continuous measurement with output $[a]$. The state $\Psi_{[a]}(q, t)$ at any instant t can be found as the solution of the time-dependent Schrödinger equation

$$i\hbar \frac{\partial \Psi_{[a]}(q, t)}{\partial t} = H_{\text{eff}} \Psi_{[a]}(q, t) \qquad (5.53)$$

with an effective Hamiltonian H_{eff} corresponding to the Lagrangian L_{eff} in (5.52) and with the choice $\Psi_{[a]}(q, 0) = \psi_0(q)$.

The time dependence of the effective Hamiltonian H_{eff} makes analytical calculations difficult. In this case one may more simply evaluate $P_{[a]}$ through the Feynman propagator $U_{[a]}$. However, due to the quadratic nature of the measurement contribution to L_{eff}, analytical calculations are essentially restricted to the case in which $L(q, \dot{q}, t)$ is a linear oscillator. In general a numerical approach must be followed and in this case the Schrödinger formalism is more suitable. The partial differential equation (5.53) is reduced to a simple finite-difference recursive equation by choosing a proper lattice to simulate the continuous space-time (Press *et al* 1986). No particular problems arise from the time dependence and the non-Hermitian nature of the differential operator H_{eff} (Presilla 1990, Presilla *et al* 1991).

In order to understand the behaviour of a nonlinear system let us first consider a linear oscillator

$$L = \frac{m}{2} \dot{q}^2 - \frac{m\omega^2}{2} q^2.$$

In this case analytical results have been obtained, allowing a test of the numerical technique. The effective Lagrangian corresponds to a forced linear oscillator

$$L_{\text{eff}} = \frac{m}{2} \dot{q}^2 - \frac{m\tilde{\omega}^2}{2} q^2 - \frac{4i\hbar}{T\Delta a^2} a(t) q + \frac{2i\hbar}{T\Delta a^2} a(t)^2$$

with renormalized complex frequency

$$\tilde{\omega}^2 = \omega^2 - \frac{4i\hbar}{mT\Delta a^2}.$$

For any choice of the measurement output $[a]$ the propagating kernel $U_{[a]}$ can be easily calculated (Feynman and Hibbs 1965, see also Chapter 3 of this book). Let us consider what happens when the quantum system is in the ground state of the unmeasured oscillator before and after the period T of the continuous measurement

$$\psi_0(q) = \psi_1(q) = \left(\frac{m\omega}{\pi\hbar}\right)^{1/4} \exp\left(-\frac{m\omega}{2\hbar}q^2\right). \tag{5.54}$$

Due to the shape of ψ_0 and ψ_1 it is natural to choose the null boundary conditions $a(0) = a(T) = 0$ for the measurement output $[a]$. Any such function $a(t)$ can be written as a Fourier sine series. We consider only measurement outputs of the form

$$a(t) = a_0 \sin \Omega t, \qquad \Omega = \frac{n\pi}{T} \tag{5.55}$$

where n is an integer. For a fixed Ω the probability distribution $P_{[a]}$ is reduced to a function of the amplitude a_0. $P(a_0)$ is a Gaussian function of width Δa_{eff},

$$P(a_0) = \sqrt{\frac{2}{\pi\Delta a_{\text{eff}}}} \exp\left(-\frac{2a_0^2}{\Delta a_{\text{eff}}^2}\right)$$

where

$$\Delta a_{\text{eff}}^{-2} = 2\,\text{Re}\left\{\frac{1}{\Delta a^2}\left(1 - \frac{4i\hbar}{mT\Delta a^2(\Omega^2 - \tilde{\omega}^2)}\right)\right.$$
$$\left. - \frac{16\hbar\Omega^2}{m\omega T^2\Delta a^4(\Omega^2 - \tilde{\omega}^2)^2}\left[1 - i\frac{\tilde{\omega}}{\omega}\left(\cot(\tilde{\omega}T) + \frac{(-1)^n}{\sin(\tilde{\omega}T)}\right)\right]^{-1}\right\}.$$

The meaning of Δa_{eff} is linked to the role of the measurement device during the evolution of the system under measurement. When the instrument error Δa is large in comparison with the characteristic quantum scale of the system under measurement, $\sqrt{\hbar/m\omega}$ in this case, the corridor of width $2\Delta a$ around the path $[a]$ contains the classical trajectory (with null boundary conditions). Thus the classical limit is

$$\Delta a_{\text{class}} = \lim_{\Delta a \to \infty} \Delta a_{\text{eff}} = \Delta a. \tag{5.56}$$

On the other hand, quantum noise arises when Δa becomes small. Also corridors far from the classical trajectory are probable. The quantum limit

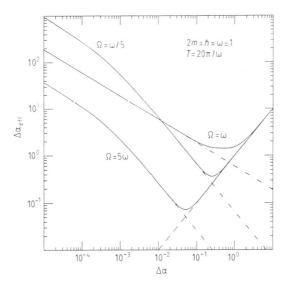

Figure 5.2: Effective uncertainty versus instrumental uncertainty for three different values of the ratio Ω/ω in the case of a linear oscillator. The broken line is the classical limit Δa_{class}; the chain lines are the quantum limits Δa_{quant}. The values of the other parameters are also indicated.

is written as

$$\Delta a_{\text{quant}} = \lim_{\Delta a \to 0} \Delta a_{\text{eff}}$$

$$= \left[\sqrt{2} \left(\frac{m}{\hbar} \right)^{3/2} T^{1/2} \Omega^2 \Delta a + \left(\frac{mT}{2\hbar} \right)^2 (\Omega^2 - \omega^2)^2 \Delta a^2 \right]^{-1/2} . (5.57)$$

In an intermediate situation Δa_{eff} interpolates between these two limits, always maintaining values larger than the instrumental error. In figure 5.2 the behaviour of Δa_{eff} versus Δa is shown for three different values of the ratio Ω/ω. As shown in (5.56) and (5.57), while the classical limit is the same for all situations, different behaviours appear in the quantum regime. The minimum effective uncertainty is maximum at the resonance condition $\Omega = \omega$. In this case it is always $\Delta a_{\text{eff}} \geq \sqrt{\hbar/m\omega}$. When $\Omega \neq \omega$ the minimum effective uncertainty estimated by the intersection between quantum and classical limits decreases as $|\Omega^2 - \omega^2|^{-1}$.

A comparison between analytical and numerical results for Δa_{eff} is shown in figure 5.3 for two different choices of the measurement parameters. Also shown are classical and quantum behaviours, corresponding to the limits expressed in (5.56) and (5.57). The difference between the numerical and analytical results is less than 0.1% and it can be further

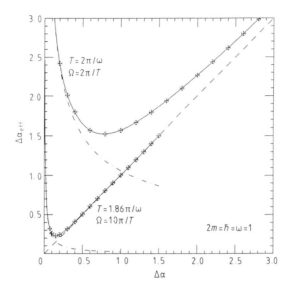

Figure 5.3: Numerical (points) and analytical (full curve) effective uncertainty versus instrumental uncertainty for a linear oscillator. Asymptotic classical (broken line) and quantum (chain curve) behaviours are also shown.

reduced by choosing higher resolution space-time lattices.

The numerical accuracy estimated above allows us to perform meaningful computations for a nonlinear oscillator represented by the Lagrangian

$$L = \frac{m}{2}\dot{q}^2 - \frac{m\omega^2}{2}q^2 - \frac{\beta}{4}q^4. \tag{5.58}$$

For comparison with the previous linear case we have chosen the initial and final states of the form (5.54) and the measurement output of the form (5.55).

In this case we expect non-Gaussian behaviour for both the wavefunction at $t = T$ and the propagator. This also means that the distribution $P(a_0)$ is not a Gaussian function. An equivalent width for $P(a_0)$ may be introduced through the definition

$$\Delta a_{\text{eff}} = \frac{1}{\sqrt{2\pi}P(0)} \int_{-\infty}^{+\infty} P(a_0)\,\mathrm{d}a_0. \tag{5.59}$$

The computed Δa_{eff} versus Δa are shown in figure 5.4 for two different values of the nonlinearity coefficient β. The comparison with the linear situation having the same parameters is also shown. It appears that the

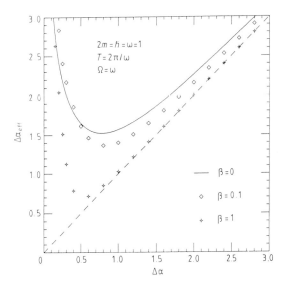

Figure 5.4: Numerical (dots) effective uncertainty versus instrumental uncertainty for the nonlinear oscillator at two different values of β. The full curve is the behaviour of the corresponding linear oscillator obtained for $\beta = 0$. The broken line represents the classical limit.

nonlinear term tends to enlarge the region in which the classical approximation is meaningful. The quartic term concentrates the final wavefunction near the result of the measurement, and this implies that the effect of quantum noise, toward a spreading of the most probable paths, is reduced.

In all the previous considerations we have chosen to deal with a Δa that is time independent. A more general class of continuous measurements is obtained by considering time dependence of Δa. In particular, to recover chaotic dynamics through quantum measurements a particular condition on the kind of measurement must be satisfied, namely the measurement process has to be a quantum nondemolition under monitoring (Adachi et al 1989). Quantum nondemolition strategies can also be analysed in the framework of the path-integral approach to continuous measurements (see Chapter 6).

6

Continuous Quantum Nondemolition Measurements

In Chapters 4 and 5 we considered typical examples of continuous measurements and saw that in each of them an optimal level of measurement precision exists. In a measurement more precise than this optimal one less information is extracted from the measurement output because of quantum measurement noise. Thus some absolute quantum restriction arises on information provided by the given type of measurement.

However, there is a class of continuous measurements for which this is not valid. No optimal level of measurement precision exists for them so that precision may be improved without limit, leading to more and more information being obtained from the measurement. No absolute quantum restriction arises for such measurements, and this is their main advantage. These evidently important measurements have been called quantum nondemolition (QND) because reduction (collapse) of the quantum state during the measurement does not destroy that part of the information about the state that we are interested in.

In this chapter this type of measurement will be considered in the framework of the path-integral method. We shall begin from the simplest case of measurement of the linear momentum of a free particle, which allows one to see all the typical features characteristic of QND measurements. A more general variety of measurements of this type will then be considered.

6.1 QUANTUM NONDEMOLITION MEASUREMENTS

Each of the continuous measurements considered in the previous chapters leads to an absolute limit on the measurability of the corresponding observable. Such limits may be formulated in different ways. One of the most important is a restriction on the measurability of force with the help of a measurement of the given type. We saw that the force acting on the measured system can be estimated from continuous measurements with some finite precision. The error of the force estimation δF depends of course on the error of measurement (instrument error) Δa. However, it cannot be made less than some characteristic value δF_{opt}. If we achieve $\delta F = \delta F_{\text{opt}}$ by reducing Δa to some optimal level Δa_{opt} and then reduce Δa further, then δF again increases so that the level δF_{opt} cannot be overcome. An absolute limit resulting from this is called usually the *standard quantum limit* (see section 5.3.3 of Chapter 5).

The picture drawn above is the result of the unavoidable back reaction of the measuring device on the measured system. This back reaction is a characteristic (and the most exciting) feature of quantum theory. It is expressed, for example, in the *Heisenberg uncertainty relation* $\Delta q\, \Delta p \gtrsim \hbar$ and means that a precise measurement of the coordinate q leads to an increase in the uncertainty of the linear momentum p.[1] This is the fundamental reason for the absolute quantum limit in continuous monitoring of the coordinate. Indeed, measurement of the coordinate disturbs the linear momentum, and this has an influence on the further dynamics of the coordinate and thus on the results of further measurements of it. Therefore, very precise measurement (monitoring) of the coordinate leads to a perturbation of the system's evolution. This is why too precise a measurement does not give much information about the system. The measurement outputs in this case become somewhat arbitrary, depending not on initial conditions and evolution laws but rather on random quantum fluctuations (quantum measurement noise) emerging in the process of measurement.

Of course, this was unpleasant for those people who tried to make precise measurements. Braginsky (1967) discovered the standard quantum limit. Braginsky, Vorontsov and Khalili (see Braginsky *et al* 1977) were the first to ask whether this difficulty is unavoidable or whether there are ways to overcome it and extract the necessary information about a quantum system with an arbitrarily low error. The result of this was the concept of *quantum nondemolition* (QND) *measurement*, the theory of which has been developed further by Braginsky *et al* (1978), Unruh (1979), Caves *et al* (1980), Caves (1983) and many others. The path-integral approach has been applied to continuous QND measurements by Golubtsova and Mensky (1989).

The idea of QND measurement is to measure a variable such that the unavoidable disturbance of the complementary one does not disturb the evolution of the chosen variable. If we measure the coordinate q, then the momentum p is disturbed. But p has an influence on the evolution of q. This is why the measurement of q at later instants will be unpredictably disturbed by its measurement at earlier instants. This is 'quantum demolition' (QD) measurement. The picture will be quite different, however, if

---

[1] See Martens and de Muynck (1990) for a contemporary and mathematically strict discussion of different uncertainty relations.

we measure the linear momentum p of a free particle. Of course this, due to the Heisenberg uncertainty relation, will disturb the coordinate q. However, this does not influence the evolution of the momentum p. Indeed, the evolution of p is trivial for a free particle: linear momentum p is constant (the conservation law holds for the momentum of a free particle). Thus measurement of p, though disturbing q, does not disturb the momentum p at later instants. Thus future measurements of p are not influenced by the preceding measurements (of course only for a free particle). This is the simplest example of QND measurement. Later we shall consider this in more detail (section 6.2).

In a more general case we have two complementary variables X, Y obeying the uncertainty relation $\Delta X \, \Delta Y \gtrsim \hbar$ and measure one of them, say X. This disturbs Y. However, the variables may be chosen in such a way that the evolution of X does not depend on Y. Therefore disturbance of Y does not influence the future values of X. As a result any preceding measurement of X has no influence on its future measurements. The measurement of X is therefore quantum nondemolition (QND). The variable X is called in this case a *quantum nondemolition variable*. We shall consider an example of this type later, choosing a quadrature component of a harmonic oscillator as a QND variable (section 6.4, see also section 9.6 of Chapter 9).

An effective formal criterion for the given variable A to be a QND variable can be formulated in terms of the Heisenberg representation of the variable. For any quantum observable (operator in the state space) A its Heisenberg representation can be defined with the help of the system Hamiltonian H and its evolution operator, which (in the case of a Hamiltonian not depending explicitly on time) has the form $U_t = \exp[(i/\hbar)Ht]$. Then the Heisenberg representation $A_H(t)$ of A is equal to

$$A_H(t) = U_t^\dagger A U_t = e^{\frac{i}{\hbar}Ht} A e^{-\frac{i}{\hbar}Ht}.$$

Then A can be proved to be a QND variable if the commutation relation

$$[A_H(t'), A_H(t'')] = 0 \tag{6.1}$$

is fulfilled for any instants t', t''.[2]

Consider for example a free particle

$$H(p,q) = \frac{p^2}{2m}. \tag{6.2}$$

The coordinate q can easily be shown to be quantum demolition because the Heisenberg operators of a coordinate, $q_H(t')$, $q_H(t'')$ do not generally

---

[2] If we are interested only in measurements performed at a discrete sequence of instants $t_1, t_2, \ldots$, then the QND character of these measurements is provided by the requirement $[A_H(t_i), A_H(t_j)] = 0$ for all instants of this sequence.

commute for $t' \neq t''$. The situation is different, however, for the linear momentum p. In this case $p_H(t) = p$ does not depend on time (this is in fact a consequence of the conservation law). Therefore the condition (6.1) is fulfilled for the Heisenberg operator of the momentum, $p_H(t)$. The momentum of a free particle is therefore a QND variable.

Considering a harmonic oscillator,

$$H = \frac{p^2}{2m} + \frac{m\omega^2 q^2}{2}, \qquad (6.3)$$

we can prove that each of two so-called quadrature components,

$$
\begin{aligned}
X &= q\cos\omega t - \frac{p}{m\omega}\sin\omega t \\
Y &= q\sin\omega t + \frac{p}{m\omega}\cos\omega t
\end{aligned}
\qquad (6.4)
$$

is a QND variable. This has been established by Thorne et al (1978) (see also Unruh 1979, Caves et al 1980, Braginsky et al 1980).

Consider now the monitoring of a QND variable in the framework of the path-integral approach, beginning from the simplest case of the linear momentum of a free (one-dimensional) particle.

6.2 MONITORING OF LINEAR MOMENTUM

Consider the simplest one-dimensional system, a free particle (6.2), and the monitoring of its linear momentum, p (see Mensky 1992c). According to the procedure developed in Chapter 4 we should present the propagator of the system in the form of a path integral and then construct the measurement amplitude, introducing the corresponding weight functional in the integrand of this path integral. However, the weight functional should now depend on the linear momentum p. Therefore the coordinate representation of a path integral is inconvenient for our task. Instead of this we may take a phase-space (Hamiltonian) form of the path integral as our starting point (for convenience we fix the time interval to be $[0, T]$):

$$U(q', q) = \int_q^{q'} d[q] \int d[p] \exp\left(\frac{i}{\hbar}\int_0^T (p\dot{q} - H(p, q))\, dt\right).$$

Introducing the weight functional we obtain the measurement amplitude:

$$U_\alpha(q', q) = \int_q^{q'} d[q] \int d[p] \exp\left(\frac{i}{\hbar}\int_0^T (p\dot{q} - H(p, q))\, dt\right) w_\alpha[p, q]. \quad (6.5)$$

The form (6.5) is appropriate for describing any continuous measurement giving information about the coordinate and linear momentum as functions of time. For monitoring of the linear momentum p the output of the measurement can be expressed by some function $b(t)$. Its meaning is that the value of the momentum p at the instant t is close to $b(t)$. If this is valid for all times in the interval $[0, T]$, the path $[p]$ in the space of momenta should be close to the path $[b]$. This means that we have to introduce the weight functional $w_{[b]}[p]$ such that it is almost equal to unity for $[p]$ close to $[b]$ and is almost zero for $[p]$ far from $[b]$. The resulting form of the measurement amplitude describing momentum monitoring is as follows:

$$U_{[b]}(q', q) = \int_q^{q'} d[q] \int d[p] \exp\left(\frac{i}{\hbar} \int_0^T (p\dot{q} - H(p, q))\, dt\right) w_{[b]}[p]. \quad (6.6)$$

A possible (and convenient) form of the functional is Gaussian:

$$w_{[b]}[p] = \exp\left(-\frac{\langle(p - b)^2\rangle}{\Delta b^2}\right) \quad (6.7)$$

where Δb is the error of measurement of the momentum and the following notation is used:

$$\langle f \rangle = \frac{1}{T} \int_0^T f(t)\, dt.$$

The measurement amplitude (6.6) now takes the form

$$U_{[b]}(q', q) = \int_q^{q'} d[q] \int d[p] \exp\left[\frac{i}{\hbar} \int_0^T dt \left(p\dot{q} - \frac{p^2}{2m}\right) - \frac{(p - b)^2}{T\Delta b^2}\right]. \quad (6.8)$$

The path integral (6.8) is of Gaussian type, and therefore it can be calculated precisely. This gives

$$U_{[b]}(q', q) = \exp\left(-\frac{\langle b^2 \rangle}{\Delta b^2} + \frac{\left(\frac{\langle b \rangle}{\Delta b^2} + \frac{i}{2\hbar}(q' - q)\right)^2}{\frac{1}{\Delta b^2} + i\frac{T}{2\hbar m}}\right). \quad (6.9)$$

Taking the square modulus, one has for the probability distribution

$$P_{[b]}(q', q) = \exp\left(-2\frac{\langle(b - \langle b \rangle)^2\rangle}{\Delta b^2} - 2\frac{\left(\langle b \rangle - \frac{m(q' - q)}{T}\right)^2}{\delta b^2}\right) \quad (6.10)$$

where

$$\delta b = \sqrt{\Delta b^2 + \frac{4\hbar^2 m^2}{T^2 \Delta b^2}}. \quad (6.11)$$

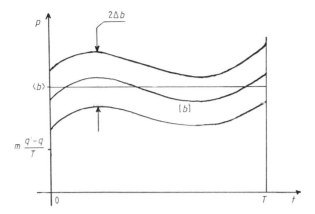

Figure 6.1: Monitoring of the momentum gives a corridor which is horizontal with accuracy Δb and is in accordance with the boundary conditions q, q' with accuracy δb.

This expression has an evident interpretation. The first term in the exponent provides that the function $b(t)$ is close to the constant function, equal to the number

$$\langle b \rangle = \frac{1}{T} \int_0^T b(t) \, dt.$$

The mean square deviation of the function $b(t)$ from $\langle b \rangle$ cannot be more than the measurement error Δb. The function $b(t)$ describes the middle line of the corridor arising as the measurement output. Therefore, this corridor should be almost horizontal, deviating from horizontal by no more than Δb.

The second term in the exponent provides that the quantity $\langle b \rangle$ is close to $m(q' - q)/T$. This means that continuous monitoring of the linear momentum is in accordance (up to the accuracy δb) with the measurement of the coordinates at the instants $t = 0, T$.

The mentioned characteristics of the distribution (6.10) can be illustrated by figure 6.1. The difference between the mean measured momentum $\langle b \rangle$ and the estimation of the momentum by coordinate measurements, $m(q' - q)/T$, is no more than δb. The horizontal line at the level $\langle b \rangle$ should lie inside the corridor having a middle line $[b]$ and width $2\Delta b$. This means that the corridor is horizontal with accuracy Δb.

The fact that the corridor is close to horizontal reflects the fact that the measured variable p is a constant of motion. The QND character of the momentum is expressed in the fact that the deviation of the middle line of the corridor from the horizontal line is less than Δb, and thus can be made arbitrarily small by choosing a sufficiently precise measuring device. No absolute quantum restriction arises in this case—a characteristic feature of QND measurement. In fact, the whole picture of momentum monitoring is essentially classical (but some quantum features are connected with the value $\langle b \rangle$, to be discussed below).

Of course, if the particle is acted upon by some external force $F(t)$, monitoring of the momentum gives a non-horizontal corridor. The deviation of the corridor from the horizontal characterizes the force. Monitoring of the linear momentum therefore allows the force to be estimated. The more accurate the monitoring, the higher is the precision of this estimation.

Consider the formula (6.11) in more detail.

In the classical regime, $\Delta b \gg \Delta b_{\mathrm{opt}}$, where

$$\Delta b_{\mathrm{opt}} = \sqrt{\frac{2\hbar m}{T}}$$

one has $\delta b = \Delta b$ so that the typical difference between the mean momentum $\langle b \rangle$ and the velocity (multiplied by the mass) estimated by the boundary conditions, $m(q' - q)/T$, is equal to

$$\left| \langle b \rangle - m \frac{q' - q}{T} \right| \simeq \Delta b$$

as might be expected from classical theory. In the quantum regime ($\Delta b \ll \Delta b_{\mathrm{opt}}$) the formula (6.11) takes the form

$$\delta b = \frac{2\hbar m}{T \Delta b}.$$

Choosing $\Delta b = \Delta b_{\mathrm{opt}}$, one can make δb minimum and equal to

$$\delta b_{\mathrm{min}} = 2\sqrt{\frac{\hbar m}{T}}. \tag{6.12}$$

Any other choice of Δb makes the accord between $\langle b \rangle$ and $m(q' - q)/T$ worse because in all regimes the following inequality is valid:

$$\Delta b \, \delta b \gtrsim \frac{2m\hbar}{T}.$$

This inequality has the form of Heisenberg's uncertainty principle. Indeed, here Δb is the measurement error of momentum monitoring and $\delta b/m$ is the uncertainty in the estimation of velocity from the outputs q, q' of the instantaneous measurement of position before and after momentum monitoring. Then $(\delta b/m)T$ is the uncertainty in position expressed through velocity and time.

We remarked above that the monitoring of linear momentum (like any QND measurement) has a purely classical character. Why then does the absolute limit (6.12) arise? The reason is that, besides momentum monitoring, we consider in fact the instantaneous measurement of position before and after monitoring, and this measurement is quantum demolition. Let us compare this with the situation when no quantum demolition measurement is performed.

Consider the momentum monitoring under the condition that the system is in the state with a definite momentum p_0 before monitoring:

$$\psi_0(q) = e^{\frac{i}{\hbar} p_0 q}.$$

Using the probability amplitude (6.9), one can find the state after the monitoring:

$$
\begin{aligned}
\psi_T(q') &= \int U_{[b]}(q', q)\psi_0(q)\, dq \\
&= \exp\left(-\frac{\langle(b - \langle b\rangle)^2\rangle}{\Delta b^2} - \frac{(\langle b\rangle - p_0)^2}{\Delta b^2}\right) e^{\frac{i}{\hbar} p_0 q}.
\end{aligned}
\tag{6.13}
$$

We see that the final state is characterized by the same momentum p_0.

We kept in equation (6.13) the numerical factor depending on the measurement result $[b]$. This factor means that the norm of the new state is less than the norm of the original one. This is because different channels of evolution are possible corresponding to different measurement outputs. This factor is nothing but a probability density for the measurement output $[b]$:

$$
P_{[b]} = \exp\left(-2\frac{\langle(b - \langle b\rangle)^2\rangle}{\Delta b^2} - 2\frac{(\langle b\rangle - p_0)^2}{\Delta b^2}\right).
\tag{6.14}
$$

This distribution is similar to (6.10), but the mean measured momentum $\langle b\rangle$ is now close to the initial momentum p_0. The possible deviation is now Δb (instead of δb in the case of a coordinate fixed before and after monitoring) and may be made arbitrarily small.

At first glance the result derived above contradicts the known uncertainty relation on the measurement of linear momentum (Landau and Lifshits 1958)

$$\Delta p\, \Delta t\, |v'' - v'| \simeq \hbar \tag{6.15}$$

where Δp is the error and Δt is the time interval of the measurement, and v', v'' are the velocities before and after the measurement. Indeed, continuous monitoring can be presented as a series of N measurements of momentum, each performed during a very short time interval $\Delta t = T/N$. Continuous monitoring must be defined as the limit of such a series of measurement for $N \to \infty$, so that $\Delta t \to 0$. Then if the errors of separate measurements Δp remain finite, $v'' - v'$ (and therefore $p'' - p'$) becomes infinite so that no conservation of linear momentum and no horizontal line as a result of measurement should be expected.

In this argument all is correct but one possibility has been omitted. We mean the case when $\Delta p \rightarrow \infty$ simultaneously with $\Delta t \rightarrow 0$. In this case $p'' - p'$ may not only remain finite but may even tend to zero. In the case $p'' - p' \rightarrow 0$ conservation of linear momentum remains valid even for a system subject to continuous monitoring of this observable. This is the case described above in the framework of the phenomenological path-integral method. This is why the function $b(t)$ obtained in this measurement is close to a horizontal line.

It is seen from this argument that $\Delta p \rightarrow \infty$ is necessary for conservation of linear momentum to be maintained. Therefore the value of momentum is not fixed in such a measurement. The result of such a measurement should be a horizontal line, but it is completely unknown at what height this line lies. Any of an infinite set of parallel horizontal lines could arise as a result of the measurement.

The formula (6.10) and its analysis given above support these conclusions. Indeed, the resulting function $b(t)$ is necessarily close to constant, and the smaller is Δb, the smaller is the deviaton of this function from constant. However, the common level $\langle b \rangle$ of this function is fixed with less accuracy δb. Moreover, this value is defined not by linear momentum monitoring, but by boundary conditions q, q', that is by instantaneous measurements of position before and after momentum monitoring. The monitoring itself only confirms that linear momentum is conserved. It gives of course some value for the conserved momentum, but this value can be arbitrary provided it is compatible with the boundary conditions (boundary measurements).

In the second considered case, when the linear momentum p_0 before monitoring is known (measured beforehand), the value $\langle b \rangle$ is fixed with the same precision Δb as the precision of the momentum conservation. This is a consequence of the QND character of the measurement of linear momentum. Of course, in this case the measurement of linear momentum before monitoring is not instantaneous. On the contrary, to obtain quite a precise value p_0 this measurement should be infinitely long, according to the uncertainty relation (6.15).

One may now doubt whether the monitoring of momentum makes any sense if it can only confirm its conservation but cannot give the value $\langle b \rangle$. It is easy to see that monitoring makes sense for estimating an external force acting on the particle. We have considered in detail only the case of null external force, when the linear momentum is conserved. If a non-zero force $F(t)$ acts, it results in a deviation from the conservation law. The monitoring of momentum then gives some function $b(t)$ differing from constant by an amount greater than the measurement error Δb. The force acting on the system can then be estimated as follows:

$$F_{est} = \frac{db(t)}{dt}.$$ (6.16)

It is just the difference of the function $b(t)$ from constant, but not its time-average $\langle b \rangle$, that is important for this estimation.

The case of non-zero external force can easily be considered in the framework of the path-integral approach. Then the formula (6.16) can be derived strictly.

6.3 VELOCITY VERSUS MOMENTUM

The weight factor (6.7) describes monitoring of the momentum, which for a free particle (6.2) turns out to be a QND variable. This can be seen from the formula (6.10) for the probability distribution of the measurement results. The first term in the exponent of (6.10) shows that the corridor resulting from the measurement is close to a horizontal line with the precision equal to the width of the corridor, Δb. From the deviation of the form of the corridor from the horizontal the experimenter can judge the force acting on the particle. Δb can be made arbitrarily small. This means that the precision of estimation of the momentum p and of the force F can be arbitrarily small too. This conclusion is typical for a QND measurement.

Take now the weight factor

$$w_{[c]}[p, q] = \exp\left(-\frac{\langle (\dot{q} - c)^2 \rangle}{\Delta c^2}\right).$$ (6.17)

It is obvious that this functional describes the monitoring of the velocity $\dot{q}$, giving the function $c(t)$ as a result. The path integral (6.5) can be calculated in this case to give the amplitude

$$U_{[c]}(q', q) = \exp\left[\frac{\langle c^2 \rangle}{-\Delta c^2 + i\frac{2\hbar}{mT}} \right.$$

$$\left. - \left(\frac{1}{\Delta c^2} + i\frac{mT}{2\hbar}\right)\left(\frac{q' - q}{T} - \frac{\langle c \rangle}{1 + i\frac{mT\Delta c^2}{2\hbar}}\right)^2 \right]$$ (6.18)

and the probability density

$$P_{[c]}(q', q) = \exp\left(-2\frac{\langle (c - \langle c \rangle)^2 \rangle}{\delta c} - 2\frac{\left(\langle c \rangle - \frac{q' - q}{T}\right)^2}{\Delta c^2}\right)$$ (6.19)

where

$$\delta c = \sqrt{\Delta c^2 + \frac{4\hbar^2}{m^2 T^2 \Delta c^2}}. \tag{6.20}$$

Now the situation is radically different from that of measuring the momentum, since the present type of measurement gives an absolute quantum limit. The first term in the exponent of (6.19) shows that the corridor for the velocity ought to be horizontal only with the precision δc. This quantity is greater than Δc and cannot be made less than

$$\delta c_{\min} = 2\sqrt{\frac{\hbar}{mT}}. \tag{6.21}$$

This is an absolute limit on the measurability of the velocity, coinciding in fact with the standard quantum limit (SQL).

The question arises naturally of the reason for such a radical difference between the momentum (as a QND variable) and the velocity (a quantum demolition (QD) variable). This difference seems strange and even paradoxical because in most cases these variables differ only by the factor m. The above consideration shows that this is not always valid in quantum mechanics.

The first explanation is that p is equal to $m\dot{q}$ only as a consequence of the classical equation of motion, but the path integral (6.5) includes all paths, even those that do not satisfy the equation. Let us be a bit more precise. The equality $p = m\dot{q}$ arises in quantum mechanics too, as an equality between quantum observables (operators). This arises when ordinary quantum dynamics takes place. In the language of paths this dynamics is expressed by an ordinary Feynman path integral including integration over all paths. However, we are dealing with a sort of modified dynamics arising under the influence of a measuring device (and the resulting state reduction). This modified dynamics is expressed by a restricted path integral. Now we see that the usual relation between linear momentum and velocity is violated in the modified dynamics. From the physical point of view this is a consequence of measurement, of the influence of a measuring device.

Let us try to make the latter affirmation more evident. To do this we shall consider a measuring device explicitly, in the framework of a simple model.

Consider a subsidiary system (called usually a probe, or a 'meter') interacting with the system under consideration. Let us suggest that the meter undergoes measuring, for indirect measuring of the main system. It was shown by Caves et al (1980) that the velocity of a particle can be measured with the help of the meter (a one-dimensional system with coordinate Q) interacting with a particle according to the following Lagrangian:

$$L_v(q, Q, \dot{q}, \dot{Q}) = \left(\frac{m}{2}\dot{q}^2 + Fq\right) - g\dot{q}Q + L_M(Q, \dot{Q}) \tag{6.22}$$

where L_M is the Lagrangian of the meter. The coefficient g here plays the role of a coupling constant. Increasing this coefficient leads to a more precise measurement.

Analogously for the (arbitrarily precise) measurement of the momentum the following Hamiltonian is needed:

$$H_p(p, P, q, Q) = \left(\frac{1}{2m}p^2 - Fq \right) + gpQ + H_M(P, Q) \qquad (6.23)$$

(it is natural to describe the measurement of momentum in Hamiltonian rather than Lagrangian formalism). Again the precision of measurement depends on the coupling constant g.

We can easily find a Lagrangian corresponding to the Hamiltonian (6.23) with the help of the formula

$$H(p, P, q, Q) = p\dot{q} + P\dot{Q} - L(q, Q, \dot{q}, \dot{Q}).$$

This gives

$$L_p = \left(\frac{m}{2}\dot{q}^2 + Fq \right) - g\dot{q}Q + \left(L_M(Q, \dot{Q}) + \frac{1}{2}g^2Q^2 \right) \qquad (6.24)$$

where L_M and H_M are connected by the relation

$$H_M(P, Q) = P\dot{Q} - L_M(Q, \dot{Q}).$$

Therefore, measurement of the momentum p is described by the inter-action Hamiltonian $H_{int} = gpQ$ where measurement of the velocity $\dot{q}$ is described by the interaction Lagrangian $L_{int} = -g\dot{q}Q$. Comparing the Lagrangians (6.22) and (6.24) gives

$$L_p - L_v = \frac{1}{2}g^2Q^2. \qquad (6.25)$$

This means that two types of measurement differ by the additional (nega-tive) elasticity of the meter.

It has been shown by Caves et al (1980) that inclusion of negative elas-ticity allows one to change measurement of the QD variable (the velocity) into measurement of the QND variable (the momentum). Any QND vari-able can be measured with arbitrarily high precision. The question arises naturally of whether transition to QND measurement allows one to evalu-ate arbitrarily precisely an internal state of the system under investigation. The answer is no. Let us consider this point in some detail.

The velocity $\dot{q}$, directly expressed through the coordinate q, is a kine-matic variable for the system under consideration. Measurement of the velocity gives direct information about an internal state of the system. However, measurement of the velocity is restricted by an absolute limit (the standard quantum limit). Thus an internal state of the system cannot be estimated with arbitrarily high precision by measurement of the velocity. What about measurement of the momentum?

The momentum p can be measured aribitrarily precisely, but it is a dynamical variable and is connected with the coordinate q indirectly. Indeed, as a consequence of the Hamiltonian (6.23), the following equation is valid:

$$p = m\dot{q} - mgQ.$$

Therefore, the momentum p is not unambiguously connected with the kinematic variable $\dot{q}$. Instead of this it depends on the coordinate Q of the meter. This dependence increases with the coefficient g (i.e. with the precision of the measurement of p). Thus the measurement of p, albeit arbitrarily precise, gives no direct information about the internal state of the system. Information about the state obtained from the measurement of p is limited despite the unlimited character of the measurement of p.

Is measurement of the momentum p then useless? The answer is no. As a consequence of equation (6.23) the following equation is valid:

$$\dot{p} = F.$$

The momentum p is therefore connected with the external force F. If the aim of the measurement is estimation of this force, the measurement of the momentum is perfectly appropriate. It allows the force to be estimated with arbitrarily high precision, not restricted by an absolute quantum limit.

Thus we have considered the measurement of momentum p and velocity $\dot{q}$ of a particle in the framework of two different approaches: in a phenomenological path-integral approach where no explicit model of the measuring device is needed, and with the help of an explicit model of the 'meter' as a part of the measuring device. The two give the same results. The final conclusions from the above argument can be formulated as follows:

- If the aim of the measurement is estimation of the external force F acting on the system (a particle), then the momentum of the system should be measured. It can be measured with arbitrary precision, and the force can be estimated from the measurement output with arbitrary precision. However, measurement of the momentum gives (in the quantum regime of measurement) limited information about the internal state of the measured system.

- If the aim of the measurement is estimation of the internal state of the system during the measurement, then instead of the momentum (which is a dynamical quantity) the kinematic variable, the velocity $\dot{q}$, should be measured. Its measurement gives direct information about the state of the system. However, the measurability of $\dot{q}$ is restricted by an absolute quantum limit derived above, and thus information about the state obtained in such a way is restricted too.

- If the measurement is made with the help of a subsidiary system, a meter, then the (unlimited) measurement of the QND variable p differs by the elasticity term $\frac{1}{2}g^2Q^2$ in the Lagrangian of the meter from the (limited) measurement of the QD variable $\dot{q}$.

Of course, the above is valid only in the case of highly precise measurement, performed in the quantum regime.[3] If the measurement is rough, i.e. is performed in the classical regime, then there is no difference between the measurement of momentum and velocity. Both give (with the corresponding accuracy) information about an internal state of the system and about the force acting on the system. This can be readily seen from the formulas (6.10) and (6.19). In fact in the classical regime the classical relation $p = m\dot{q}$ is valid and the two variables practically coincide.

We have considered here only the simplest QND variable, the momentum of a free particle, and the corresponding QD variable. This is sufficient to solve those questions of principle we were interested in. More complex QND variables and their monitoring can be discussed in a quite analogous way. We shall consider shortly some technical questions of the path-integral approach to more complicated QND variables in section 6.4, though without touching questions of principle.

6.4 QND MEASUREMENTS OF OSCILLATORS

Consider now more sophisticated QND measurements in the framework of the path-integral approach. We will follow the arguments of Golubtsova and Mensky (1989), and much more detail can be found in that paper.

Let us use the formula (6.5) for the concrete case of monitoring of the observable $A = A(p, q)$. Let the result (output) of this monitoring be the path $[a]$. This means that the estimation of the function $A(t)$ is $a(t)$. Then the amplitude (6.5) takes the form

$$U_{[a]}(q', q) = \int_q^{q'} d[q] \int d[p] \exp\left(\frac{i}{\hbar} \int_0^T (p\dot{q} - H(p, q))\, dt\right) w_{[a]}[p, q].$$

(6.26)

The weight functional describing this measurement can be chosen in the Gaussian form

$$w_{[a]}[p, q] = \exp\left(-\frac{\|A - a\|^2}{\Delta a^2}\right)$$

where the notation

$$\|\xi\|^2 = \frac{1}{T} \int_0^T |\xi(t)|^2\, dt$$

---

[3] For the definition of this regime see the detailed analysis in Chapters 4 and 5.

is used.

It is evident that the amplitude (6.26) can be rewritten as

$$U_{[a]}(q', q) = \int_q^{q'} d[q] \int d[p] \exp\left(\frac{i}{\hbar} \int_0^T (p\dot{q} - \tilde{H}(p, q)) \, dt\right) \qquad (6.27)$$

where an effective Hamiltonian is introduced:

$$\tilde{H} = H - i\gamma(A - a)^2, \qquad \gamma = \frac{\hbar}{T\Delta a^2}. \qquad (6.28)$$

Consider an observable A of the form

$$A = \rho q + \sigma p \qquad (6.29)$$

where ρ and σ are real functions of time. If, in addition, the Hamiltonian H is quadratic then the path integral (6.27) is of Gaussian type and can be evaluated to give

$$U_{[a]}(q', q) = C \exp\left(\frac{i}{\hbar} \int_0^T (\tilde{p}\dot{\tilde{q}} - \tilde{H}(\tilde{p}, \tilde{q})) \, dt\right) \qquad (6.30)$$

where $\tilde{p}, \tilde{q}$ denote the solution to Hamiltonian equations for the effective Hamiltonian:

$$\dot{p} = -\frac{\partial \tilde{H}}{\partial q}, \qquad \dot{q} = \frac{\partial \tilde{H}}{\partial p}, \qquad q(0) = q, \qquad q(T) = q'. \qquad (6.31)$$

It can be shown that the probability density is then equal to

$$P_{[a]}(q', q) = |U_{[a]}(q', q)|^2 = \exp\left(-2\frac{\|\tilde{A} - a\|^2}{\Delta a^2}\right) \qquad (6.32)$$

with

$$\tilde{A} = \rho\tilde{q} + \sigma\tilde{p}.$$

The quantity (6.32) may be interpreted as a probability density that monitoring of A will give the output $[a]$. This enables one to find the variance of the measurement outputs if the force acting on the system is given. On the other hand, it enables one to solve the inverse problem and find the accuracy with which the force can be estimated from the measurement result. It is important that the formula (6.32) takes into account not only errors of the measuring device but also the variance of outputs due to the back reaction of the measurement process on the measured system. One may say that this method takes into account both classical and quantum measurement noise. It is supposed implicitly that the best of all possible measuring apparatus is used, leading to the given type of measurement.

This means that this apparatus provides all the necessary information with the minimum possible perturbation of the measured system.

It is easily seen from (6.32) that those measurement results [a] which have sufficiently large probabilities satisfy the inequality

$$\|\tilde{A} - a\|^2 \lesssim \frac{\Delta a^2}{2}. \tag{6.33}$$

Indeed, in the limits of this inequality the probability density (6.32) remains approximately (up to the order of magnitude) constant, but beyond these limits the probability density decays exponentially.

Let us apply the formulas obtained above to the analysis of continuous QND measurements. We shall restrict our consideration to the case to which these formulas are applicable, i.e. linear (in coordinates and momenta) observables (6.29) and quadratic Hamiltonians. Though we consider explicitly only one-dimensional systems, a generalization to multidimensional systems is straightforward and was in fact performed by Golubtsova and Mensky (1989).

Equations (6.31) can be put into the form

$$\dot{p} = -\frac{\partial H}{\partial q} + 2i\gamma\rho(A - a), \qquad \dot{q} = \frac{\partial H}{\partial p} - 2i\gamma\sigma(A - a). \tag{6.34}$$

Using these equations, one can easily find

$$\dot{\tilde{A}} = \left(\dot{\rho}q + \dot{\sigma}p + \rho\frac{\partial H}{\partial p} - \sigma\frac{\partial H}{\partial q} \right)\Bigg|_{p=\tilde{p}, q=\tilde{q}}. \tag{6.35}$$

Note that the terms including $\gamma = \hbar/\tau\Delta a^2$ were cancelled in (6.35), so that this equation coincides with the corresponding equation of classical theory generated by the Hamiltonian H.

Let the right-hand side of equation (6.35) be a function of $\tilde{A}$,

$$G(\tilde{p}, \tilde{q}) = \chi(\tilde{A}) \tag{6.36}$$

where

$$G(p, q) = \dot{\rho}q + \dot{\sigma}p + \rho\frac{\partial H}{\partial p} - \sigma\frac{\partial H}{\partial q}.$$

We shall see that equation (6.36) is a condition for the variable A to be QND.

In the one-dimensional case[4] the functional equation (6.36) is equivalent to the differential equation

$$\sigma\frac{\partial G}{\partial q} = \rho\frac{\partial G}{\partial p}$$

---

[4] See Golubtsova and Mensky (1989) for the multidimensional case.

or equivalently

$$\rho^2 \frac{\partial^2 H}{\partial p^2} - 2\rho\sigma \frac{\partial^2 H}{\partial p\, \partial q} + \sigma^2 \frac{\partial^2 H}{\partial q^2} = \sigma\dot{\rho} - \rho\dot{\sigma}. \qquad (6.37)$$

If equation (6.37) is fulfilled, then the equation for $\tilde{A}$ takes the form

$$\dot{\tilde{A}} = \chi(\tilde{A}). \qquad (6.38)$$

It is important that this equation has the same form as in classical theory.[5] The solution to this equation is nothing but a classical trajectory of the present observable. The inequality (6.33) can be easily interpreted in this case. It means that the measurement output (the function $a(t)$) can differ (in the sense of mean square deviation) from a classical trajectory by no more than the measurement error Δa. This is just what it should be in classical measurement theory. One can conclude therefore that *no quantum effect is important* in the case considered (when the condition (6.37) is fulfilled). The variance of the measurement results can be made arbitrarily small by reducing Δa. This is characteristic of QND measurement.

It is easy to verify that this characterization of QND measurement derived from the path integration method (in the special case of a linear observable and quadratic Hamiltonian) is in accordance with equation (6.1). Indeed, if a classical variable A satisfies equation (6.38), then the same equation is valid for the Heisenberg operator A_H:

$$\dot{A}_H = \chi(A_H).$$

The solution to this equation

$$A_H(t) = \chi_1(A_H(t_0), t_0, t)$$

shows that the variable A is contained in the class of QND variables satisfying equation (6.1) (see Caves *et al* 1980).

Thus the considered class of QND variables arises when we require the variable $\tilde{A}$ to satisfy the first-order differential equation (6.38). The other classes of QND variables correspond to differential equations of higher order.

To illustrate condition (6.37) consider a harmonic oscillator acted upon by a force:

$$H = \frac{p^2}{2m} + \frac{m\omega^2 q^2}{2} - Fq \qquad (6.39)$$

---

[5] The solution of equation (6.38) coincides with the classical trajectory for the given boundary conditions (6.31) only if this solution is real. This gives an additional relation between q, q' and $[a]$. For example, for $A - p$, $H = p^2/2m - Fq$ this relation takes the form $m(q' - q)/T = \langle a \rangle$. It can be shown that the distribution $\Pi_{[a]}(q', q)$ provides that only those measurement outputs $[a]$ are probable for which this relation is satisfied with sufficient precision.

where $F = F(t)$. In this case equation (6.37) takes the form

$$\frac{d}{dt}\left(\frac{\rho}{\sigma}\right) = \frac{1}{m}\left(\frac{\rho}{\sigma}\right)^2 + m\omega^2. \qquad (6.40)$$

This equation has a solution of the form

$$\frac{\rho}{\sigma} = m\omega \tan[\omega(t - t_0)]$$

so that we have a set of linear QND variables (in p and q)

$$A(t) = \sigma(t)[p + q\, m\omega \tan \omega(t - t_0)] \qquad (6.41)$$

satisfying the equation

$$\dot{A} - \left(\frac{1}{m}\frac{\rho}{\sigma} + \frac{\dot{\sigma}}{\sigma}\right) A = \sigma F.$$

One can verify directly that this is a QND variable. The quadrature components (6.4) are special cases of (6.41). Actually the variables of the class (6.41) differ from quadrature components only by time-dependent factors.

The condition (6.37) enables one to look for QND variables in non-trivial cases too. Consider an oscillator with time-dependent frequency $\omega = \omega(t)$ for example. In this case the condition (6.37) leads to equation (6.40), but now it is the Riccati equation. By substituting

$$\frac{\rho}{\sigma} = -m\frac{\dot{v}}{v}$$

one can reduce this equation to the equation of a classical oscillator with variable frequency:

$$\ddot{v} + \omega^2(t)v = 0.$$

If a solution

$$\frac{\rho}{\sigma} = f(t)$$

to equation (6.40) is found in some way, then a QND variable for an oscillator with time-dependent frequency has the form

$$A = \sigma(t)\big(p + q\, f(t)\big)$$

and satisfies the equation

$$\dot{\tilde{A}} = \left(\frac{\dot{\sigma}}{\sigma}\right)\tilde{A} + \sigma F.$$

We have considered here for simplicity only one-dimensional systems. However, multidimensional ones can be dealt quite analogously (see Golubtsova and Mensky 1989). In that paper models of indirect QND measurements (that is models with part of a measuring device, a meter, taken into account explicitly) are considered in greater detail.

7

Measurement of an Electromagnetic Field

Up to now we have dealt with the nonrelativistic problem of continuous measurements, i.e. measurements prolonged in time. This problem can be solved by restriction of the Feynman path integral. In this chapter we will consider quantum measurements of relativistic systems, such as quantum fields. The quantum theory of fields can be constructed with the help of a functional integral with integration over field configurations. It is natural that a description of measurements of quantum fields can be achieved by the restriction of such integrals.

The only example that will be investigated in detail is measurement of an electromagnetic field. This is of special interest because the classic papers of Landau and Peierls (1931) and Bohr and Rosenfeld (1933) were devoted to this problem. We shall see that new light can be shed on this classical area with the help of the technique of restricted functional integrals.

The main point is that different definitions of electromagnetic field strength lead to different conclusions about the existence or nonexistence of an absolute quantum restriction on the measurability of the field strength.

Other quantum fields can also be considered in the framework of the path-integral approach. Quantum measurements of a gravitational field will be mentioned in Chapter 9 in connection with the action uncertainty principle.

7.1 EXPOSITION OF THE METHOD

Consider the theory of a relativistic (generally multicomponent) field Φ with the action functional

$$S[\Phi] = \int_\Omega \mathrm{d}^4 x\, L(\Phi, \partial\Phi) \tag{7.1}$$

depending on the field configuration $[\Phi]$. Integration here is carried out over some space-time region Ω, and $\partial\Phi$ denotes partial derivatives of Φ in coordinates. The quantum dynamics of the field can be described with the help of the integral over field configurations. Consider an amplitude

$$U = \int d[\Phi] \, \exp(iS[\Phi]) \qquad (7.2)$$

with integration over all configurations of the field Φ in the region Ω satisfying the given conditions at the boundary $\partial\Omega$ of this region. The amplitude U depends therefore on the region Ω as well as on the boundary conditions imposed on Φ at the boundary $\partial\Omega$.

Note that we use natural units here (as is usual in relativistic theory), so that the velocity of light and Planck's constant are accepted as units, $c = \hbar = 1$. The final results will be formulated in ordinary units.

Amplitudes of the form (7.2) are widely used in quantum field theory. For example, if Ω is the region between two space-like surfaces Σ_1 and Σ_2, then the quantity (7.2) may be interpreted as an amplitude for propagation of the field between two instants of time corresponding to the surfaces. To be more precise, this is a probability amplitude for the field to be observed in a definite configuration at the surface Σ_2 (the configuration being determined by the boundary conditions at Σ_2) given the configuration at Σ_1 (determined by the boundary conditions at Σ_1).

The form (7.2) for the amplitude is valid in the case when no information is available about the field configuration inside the region Ω, so that all configurations are treated on an equal footing. Suppose now that some measurement is carried out in the region Ω. The result of this measurement supplies some information about the field configuration in Ω. Denote the measurement output by α and suppose that the information about the field, owing to this result, is characterized by the positive functional $w_\alpha[\Phi]$. This means that the larger $w_\alpha[\Phi]$, the more probable is the configuration $[\Phi]$, in the light of information given by the measurement output α. Then the expression (7.2) for an amplitude must be changed into

$$U_\alpha = \int d[\Phi] \, w_\alpha[\Phi] \, \exp(iS[\Phi]). \qquad (7.3)$$

In this expression integration over field configurations is carried out with the weight w_α, characterizing the probabilities of these configurations in the light of the measurement output.

The amplitude (7.3) can be used to calculate the probabilities of different measurement outputs. To do this one should consider amplitudes corresponding to the same region Ω and the same boundary conditions at $\partial\Omega$, but to different measurement outputs α. Then the quantity (7.3) can be interpreted as a probability amplitude for the measurement to give the result α. In this situation the fixed boundary conditions at $\partial\Omega$ describe additional information about the field, this information resulting from additional measurements carried out at the points of $\partial\Omega$ or, more generally, outside Ω. The squared modulus

$$P_\alpha = |U_\alpha|^2$$

then gives the probabilities (or rather probability densities) of different measurement outputs α. This gives a complete picture of the measurement.

7.2 MEASUREMENT AMPLITUDE

Let us now evaluate the measurement amplitude (7.3) in the case of measurement of the electromagnetic field strength (Mensky 1988b). The functional integral (7.2) has the following form in this case (Itzykson and Zuber 1980):

$$U = \int d[A]\, \delta(\partial_\mu A^\mu) \exp\left(-\frac{i}{4} \int d^4x\, (\partial_\mu A_\nu - \partial_\nu A_\mu)(\partial^\mu A^\nu - \partial^\nu A^\mu)\right).$$
(7.4)

Integration is carried out here over all vector potentials $A_\mu(x)$ (in the region Ω) satisfying an additional condition $\partial_\mu A^\mu = 0$. The latter is provided by the delta-functional in the integrand. This functional is included in the definition of the measure, so that the symbol $d[\Phi]$ in (7.2) should be specified as $d[A]\, \delta(\partial_\mu A^\mu)$ for the electromagnetic field. The exponent in the integrand contains the Lagrangian of the electromagnetic field. Expressing it in terms of the field strength, one has

$$U = \int d[A]\, \delta(\partial_\mu A^\mu) \exp\left(-\frac{i}{2} \int d^4x\, (\boldsymbol{H}_A^2 - \boldsymbol{E}_A^2)\right).$$
(7.5)

Here $\boldsymbol{H}_A$ and $\boldsymbol{E}_A$ denote the strength of the magnetic and electric fields corresponding to the vector potential A_μ.

The formula (7.5) is valid if no information exists on the field configuration in the region Ω. Now let the measurement of the magnetic and electric strengths be made, with corresponding errors ΔH and ΔE, at each point of the region Ω. If the result of the measurement is represented by the configurations $\vec{H}(x)$, $\vec{E}(x)$, then the resulting information about the field can be characterized by the functional

$$w_{[\vec{H}, \vec{E}]}[A] = \exp\left[-\frac{1}{\omega} \int d^4x \left(\frac{(\boldsymbol{H}_A - \vec{H})^2}{\Delta H^2} + \frac{(\boldsymbol{E}_A - \vec{E})^2}{\Delta E^3}\right)\right]$$
(7.6)

where ω is the measure (four-dimensional volume) of the region Ω. The larger the square average deviation of the field $\boldsymbol{H}_A$, $\boldsymbol{E}_A$ from the configuration $\vec{H}$, $\vec{E}$, the smaller is the functional (7.6).[1] The functional decreases by a factor e when the mean square deviation of the magnetic field becomes larger than ΔH and/or the deviation of the electric field becomes larger than ΔE.

Therefore, the functional (7.6) describes the packet of configurations that corresponds to the field measurement giving the output $[\vec{H}, \vec{E}]$. This functional may be taken as a weight functional in the formula (7.3). This gives

$$U_{[\vec{H},\vec{E}]} = \int d[A]\, \delta(\partial_\mu A^\mu) \exp\left[-\frac{i}{2} \int d^4x\, (H_A^2 - E_A^2)\right.$$
$$\left. -\frac{1}{\omega} \int d^4x\, \left(\frac{(H_A - \vec{H})^2}{\Delta H^2} + \frac{(E_A - \vec{E})^2}{\Delta E^2}\right)\right] \tag{7.7}$$

for the amplitude of the electromagnetic field measurement.

The functional integral (7.7) can be calculated with the help of the standard trick of shifting to zero boundary conditions. Let A^{class} be an extremal (classical) field configuration, satisfying a given boundary condition at $\partial\Omega$, and let $\vec{H}_{\text{class}}$, $\vec{E}_{\text{class}}$ be the corresponding field strengths. Define the vector potential B_μ by

$$A_\mu = A_\mu^{\text{class}} + B_\mu \tag{7.8}$$

and take it as a new integration variable. Then integration over B_μ should be carried out with zero boundary conditions. The integral (7.7) takes the following form:

$$U_{[\vec{H},\vec{E}]} = \exp(iS[A^{\text{class}}]) \int d[B]\, \delta(\partial_\mu B^\mu) \exp\left[-\frac{i}{2} \int d^4x\, (H_B^2 - E_B^2)\right.$$
$$\left. -\frac{1}{\omega} \int d^4x\, \left(\frac{(H_B - H')^2}{\Delta H^2} + \frac{(E_B - E')^2}{\Delta E^2}\right)\right] \tag{7.9}$$

where H', E' denote the configurations $\vec{H}$, $\vec{E}$ but 'reduced to the null boundary conditions', i.e.

$$\vec{H} = \vec{H}_{\text{class}} + H', \qquad \vec{E} = \vec{E}_{\text{class}} + E'. \tag{7.10}$$

Denote

$$\alpha = \frac{\omega \Delta H^2}{2}, \qquad \beta = \frac{\omega \Delta E^2}{2}. \tag{7.11}$$

---

[1] The actual form of the functional w_α depends on the choice of measuring device. However, the actual form is not important for estimation up to the order of magnitude, and only such an estimation is the aim of the present consideration. The functional (7.6) is sufficiently realistic for the measurement in question, and it turns out to be convenient for calculation, since it leads to Gaussian functional integrals.

Then the quadratic form in the exponent in (7.9) can be expressed as

$$H_B^2 - i\alpha^{-1}(H_B - H')^2 = (1 - i\alpha^{-1})H_C^2 + (1 + i\alpha)^{-1}H'^2 \qquad (7.12)$$

with

$$H_C = H_B - (1 + i\alpha)^{-1}H'$$

and analogously for an electric field. As a result, the integral (7.9) takes the form

$$
\begin{aligned}
U_{[\vec{H},\vec{E}]} = \exp\Big(& iS[A^{\text{class}}] \\
& - \frac{i}{2} \int d^4x \left[(1 + i\alpha)^{-1}H'^2 - (1 + i\beta)^{-1}E'^2 \right] \Big) \\
\times & \int d[B]\, \delta(\partial_\mu B^\mu) \exp\Big(-\frac{i}{2} \int d^4x \big[(1 - i\alpha^{-1})H_C^2 \\
& - (1 + i\beta^{-1})E_C^2 \big] \Big).
\end{aligned}
\qquad (7.13)
$$

Now change the integration variable B_μ to C_μ, the latter being the potential corresponding to the fields H_C, E_C. Then it is easy to see that the integral in (7.13) does not depend on the fields H', E'. Therefore this integral may be omitted (a normalization factor does not interest us). This gives the following expression for the square modulus of the measurement amplitude:

$$P_{[\vec{H},\vec{E}]} = |U_{[\vec{H},\vec{E}]}|^2 = \exp\left[-\int d^4x \left(\frac{H'^2}{\alpha + \alpha^{-1}} + \frac{E'^2}{\beta + \beta^{-1}} \right) \right]. \qquad (7.14)$$

This quantity is nothing but a probability density for the measurement to give the result expressed by the configuration $[\vec{H}, \vec{E}]$.

7.3 ANALYSIS OF THE RESULTS

With the definitions (7.10), (7.11) the probability distribution (7.14) can be expressed as follows:

$$P_{[\vec{H},\vec{E}]} = \exp\left[-\frac{2}{\omega} \int_\Omega d^4x \left(\frac{(\vec{H} - \vec{H}_{\text{class}})^2}{\Delta H^2 + \frac{4}{\omega^2 \Delta H^2}} + \frac{(\vec{E} - \vec{E}_{\text{class}})^2}{\Delta E^2 + \frac{4}{\omega^2 \Delta E^2}} \right) \right]. \qquad (7.15)$$

Thus the probability density for measurement results is found, up to the normalizing factor. The normalization is not actually necessary for us, since only the relative probabilities of different measurement results will be exploited.

The interpretation of the probability distribution (7.15) is evident. The measurement of magnetic and electric fields in the region Ω most probably results in the configuration $[\vec{H}_{\text{class}}, \vec{E}_{\text{class}}]$, defined by the boundary conditions at $\partial\Omega$. According to (7.15), the output of measurement $[\vec{H}, \vec{E}]$ may differ from the classical configuration in such a way that the mean square deviations

$$\|\vec{H} - \vec{H}_{\text{class}}\|^2 = \frac{1}{\omega} \int_{\Omega} d^4x \, (\vec{H} - \vec{H}_{\text{class}})^2$$

$$\|\vec{E} - \vec{E}_{\text{class}}\|^2 = \frac{1}{\omega} \int_{\Omega} d^4x \, (\vec{E} - \vec{E}_{\text{class}})^2 \qquad (7.16)$$

are not too large.

Indeed, it is seen from (7.15) that the probability density remains near its maximum value while the modulus of the exponent is less than unity, i.e. while the deviations (7.16) satisfy the following inequalities:

$$\|\vec{H} - \vec{H}_{\text{class}}\|^2 \lesssim \delta H^2, \qquad \|\vec{E} - \vec{E}_{\text{class}}\|^2 \lesssim \delta E^2 \qquad (7.17)$$

with

$$\delta H^2 = \Delta H^2 + \frac{\Delta H_{\text{opt}}^4}{\Delta H^2}, \qquad \delta E^2 = \Delta E^2 + \frac{\Delta E_{\text{opt}}^4}{\Delta E^2} \qquad (7.18)$$

and

$$\Delta H_{\text{opt}}^2 = \Delta H_{\text{opt}}^2 = \frac{2}{\omega}.$$

If the deviations (7.16) (or at least one of them) exceed the characteristic values (7.18), then the probability density (7.15) decays exponentially, so that the corresponding measurement results turn out to be practically improbable. In other words, the values (7.18) characterize the variance of the measurement outputs.

Consider what this dispersion is in three characteristic regimes of measurement. If

$$\Delta H \gg \Delta H_{\text{opt}}, \qquad \Delta E \gg \Delta E_{\text{opt}}$$

then the second term on the right-hand side of each of the formulas (7.18) is negligible, so that (7.18) takes the form

$$\delta H = \Delta H, \qquad \delta E = \Delta E. \qquad (7.19)$$

This means that the variance of the measurement outputs is in this case a consequence of measurement (apparatus) errors only, in complete agreement with classical theory. Therefore, under the conditions $\Delta H \gg \Delta H_{\text{opt}}$, $\Delta E \gg \Delta E_{\text{opt}}$ the classical regime of measurement is realized, with quantum effects being negligible. This regime corresponds to a sufficiently rough measurement. The variance of the measurement outputs in this regime decreases as the measurement becomes more precise.

If, on the contrary,

$$\Delta H \ll \Delta H_{\text{opt}}, \qquad \Delta E \ll \Delta E_{\text{opt}}$$

then the first terms in (7.18) become negligible, so that

$$\delta H^2 = \frac{\Delta H_{\text{opt}}^2}{\Delta H}, \qquad \delta E^2 = \frac{\Delta E_{\text{opt}}^2}{\Delta E}. \tag{7.20}$$

This is the quantum regime of measurement, and it has a paradoxical character. The value δH increases as ΔH decreases. Therefore, the more precise the measurement becomes, the more the variance of the measurement outputs increases. As a result, objective information about the field decreases. This is of course a consequence of unavoidable back reaction of the measuring device on the quantum field. The same conclusions are also valid about the characteristics δE, ΔE of the measurement of electrical field.

The optimal regime of measurement is between the classical and quantum regimes, when $\Delta H \simeq \Delta H_{\text{opt}}$, $\Delta E \simeq \Delta E_{\text{opt}}$. In this case the variance of the measurement results has the minimum possible value:

$$\delta H_{\text{min}} = \sqrt{2}\Delta H_{\text{opt}}, \qquad \delta E_{\text{min}} = \sqrt{2}\Delta E_{\text{opt}}. \tag{7.21}$$

A measurement in this regime provides the maximum objective information about the state of the field.

The last formula presents an *absolute restriction* on the precision with which the field strength at each point of Ω can be estimated by means of measurement arranged in the limits of Ω. For a region Ω of size l in space directions and of size τ in the time direction, the absolute restriction on measurability of the field turns out to be

$$\delta H_{\text{min}} = \delta E_{\text{min}} = \frac{2}{\sqrt{\tau l^3}}. \tag{7.22}$$

Up to now we have used natural units, with $c = \hbar = 1$. In ordinary units one has

$$\delta H_{\text{min}} = \delta E_{\text{min}} = 2\sqrt{\frac{\hbar}{\tau l^3}}. \tag{7.23}$$

Remark 1 It may seem strange that the volume ω of the space-time region Ω arises in the final formulas for the pointwise measurements. The reason is that measurement of the field at each point $x \in \Omega$ is carried out with the help of some sort of procedure arranged in the whole Ω.

The formula (7.23) agrees with the results of the classic paper by Landau and Peierls (1931). In that paper they derived quantum restrictions on the measurability of an electromagnetic field by considering *gedanken* experiments.

Later Bohr and Rosenfeld (1933) re-examined this question and obtained weaker restrictions. It is important that no absolute limitation on the measurability of an electromagnetic field was obtained in their work. At first glance the results of the present consideration support the results of Landau and Peierls and contradict those of Bohr and Rosenfeld. However, this is only because we did not use a certain arbitrariness in the definition of the field strength. The alternative definition will be analysed in sections 7.4, 7.5.

7.4 ANOTHER DEFINITION OF THE FIELD STRENGTH

In choosing the weight functional in the form (7.6) we accepted the definition of the field strengths as follows:

$$\begin{aligned}
\boldsymbol{H}_A &= (\partial_0 A_1 - \partial_1 A_0,\ \partial_0 A_2 - \partial_2 A_0,\ \partial_0 A_3 - \partial_3 A_0) \\
\boldsymbol{E}_A &= (\partial_2 A_3 - \partial_3 A_2,\ \partial_3 A_1 - \partial_1 A_3,\ \partial_1 A_2 - \partial_2 A_1).
\end{aligned} \qquad (7.24)$$

This definition may seem natural and even the only one possible. However, another possibility will arise if one uses, instead of (7.4), another, so-called first-order form for the action of an electromagnetic field and the corresponding form for an amplitude (Itzykson and Zuber 1980). Consider this possibility along the lines of the paper by Mensky (1989).

Choose the following formula for an amplitude:

$$U = \int d[A]\, d[F] \exp\left(i \int_\Omega d^4x\, L(A, F)\right). \qquad (7.25)$$

Here

$$L = \frac{1}{4} F_{\mu\nu} F^{\mu\nu} - \frac{1}{2} F^{\mu\nu}(\partial_\mu A_\nu - \partial_\nu A_\mu) \qquad (7.26)$$

is the first-order Lagrangian of the electromagnetic field and

$$d[A] = \delta(\partial_\mu A^\mu) \prod_{\mu,x} dA_\mu(x), \qquad d[F] = \prod_{\mu,\nu,x} dF_{\mu\nu}(x)$$

is the measure of the functional integration over the field configurations. The physical meaning of the amplitude (7.25) depends on the choice of the space-time region Ω and the conditions imposed on the field configuration $[A, F]$ on the boundary $\partial\Omega$ of this region.

To investigate the measurement procedure, let us introduce instead of (7.25) the amplitude including a weight functional:

$$U_{[\vec{H},\vec{E}]} = \int d[A]\, d[F] \exp\left(i \int_\Omega d^4x\, L(A, F)\right) w_{[\vec{H},\vec{E}]}[A, F]. \qquad (7.27)$$

Here $[\vec{H}, \vec{E}]$ is the configuration of the strengths $\vec{H}(x)$, $\vec{E}(x)$ of the electric and magnetic fields in the region Ω. This configuration represents the output of the measurement, and the functional $w_{[\vec{H},\vec{E}]}$ describes this measurement.

If we choose the functional in the form based on the definition (7.24),

$$w^A_{[\vec{H},\vec{E}]}[A, F] = \exp\left[-\frac{1}{\omega}\int d^4x \left(\frac{(\vec{H} - \boldsymbol{H_A})^2}{\Delta H^2} + \frac{(\vec{E} - \boldsymbol{E_A})^2}{\Delta E^2}\right)\right] \quad (7.28)$$

then the results of the preceding section are reproduced. However, now we may choose the functional in an alternative form (Mensky 1989):

$$w^F_{[\vec{H},\vec{E}]}[A, F] = \exp\left[-\frac{1}{\omega}\int d^4x \left(\frac{(\vec{H} - \boldsymbol{H_F})^2}{\Delta H^2} + \frac{(\vec{E} - \boldsymbol{E_F})^2}{\Delta E^2}\right)\right] \quad (7.29)$$

where another definition of field strengths is used:

$$\boldsymbol{H_F} = (F_{01}, F_{02}, F_{03}), \qquad \boldsymbol{E_F} = (F_{23}, F_{31}, F_{12}) \quad (7.30)$$

The choice (7.28) reproduces the preceding results because the functional $w^A_{[\vec{H},\vec{E}]}[A, F]$ does not actually depend on $[F]$. Therefore integration over $[F]$ can be performed explicitly and the problem can be reduced to the same integrals as in the preceding section. In particularly, an absolute restriction on the measurability of an electromagnetic field (7.21)–(7.23) of the type obtained by Landau and Peierls (1931) is predicted in such a way.

Consider now the second possibility, connected with the choice (7.29) of the measurement functional, that is with the definition of the field strength as in (7.30). To calculate the distribution of the measurement outputs in this case, one should substitute the expression (7.29) for the functional in the integral (7.27). We will not follow the details of the calculation, but give here only the final result. The distribution of measurement outputs in this case has the form

$$P^F_{[\vec{H},\vec{E}]} = \exp\left[-\frac{2}{\omega}\int_\Omega d^4x \left(\frac{(\vec{H} - \vec{H}_{\text{class}})^2}{\Delta H^2} + \frac{(\vec{E} - \vec{E}_{\text{class}})^2}{\Delta E^2}\right)\right]. \quad (7.31)$$

It can be seen from this expression that the electric field configuration $[\vec{E}]$ obtained as a result of measurement can differ from the classical configuration $[\vec{E}_{\text{class}}]$ by an amount of the order of ΔE, coinciding with the error of measurement. Similarly, the result $[\vec{H}]$ of measuring the magnetic field can differ from the classical configuration $[\vec{H}_{\text{class}}]$ by no more than ΔH. Thus the spread of the results of the measurement is determined in this case only by the error of the measuring instrument, and quantum effects are unimportant.

If the errors of the measurement ΔH, ΔE are decreasing, the spread can be made arbitrarily small. Therefore in this case there is no absolute limitation on the measurability of the field strength, in agreement with the conclusions of Bohr and Rosenfeld (1933).

We see that in the framework of quantum theory an electromagnetic field strength can be defined in two ways: in terms of the 'curl' of the potential (7.24) or in terms of the field tensor (7.30). A formal procedure for considering measurement of the field strength leads to an absolute restriction on measurability (coinciding with the results of Landau and Peierls) if the first definition is used. The same procedure leads to no absolute restriction (as in the paper by Bohr and Rosenfeld) if the second definition is used. We shall attempt to show now that different definitions correspond to different types of measurement, and each has its own advantage.

7.5 EXPLICIT MODEL OF A METER

To analyse the measurement process in quantum mechanics one can use a simple model of the sensitive element of the instrument, sometimes called a 'meter'. The purely classical part of the instrument is not then explicitly considered. The coupling of the apparatus to the measured system is described by an interaction Lagrangian that is proportional to the product of the measured quantity and the coordinate of the meter. In the case we are interested in, the measured quantity is the field, and therefore the role of the coordinate of the meter must also be played by some field. We denote it by $G^{\mu\nu}(x)$. Let the Lagrangian of the meter be $L_G(G)$. Then the total Lagrangian describing the measurement has the form

$$L_A(A, F, G) = L(A, F) + g\,(\partial_\mu A_\nu - \partial_\nu A_\mu)G^{\mu\nu} + L_G(G) \qquad (7.32)$$

if we measure the 'curl' of the potential, and

$$L_F(A, F, G) = L(A, F) + g\,F_{\mu\nu}G^{\mu\nu} + L_G(G) \qquad (7.33)$$

if the field tensor is measured.

To compare the two schemes of measurement, we go over from the Lagrangians (7.32) and (7.33) to equivalent Lagrangians containing only the potential A_μ and the field $G^{\mu\nu}$ but not the tensor $F_{\mu\nu}$:

$$
\begin{aligned}
L_A(A, G) &= L(A) + g\,(\partial_\mu A_\nu - \partial_\nu A_\mu)G^{\mu\nu} + L_G(G) & (7.34)\\
L_F(A, G) &= L(A) + g\,(\partial_\mu A_\nu - \partial_\nu A_\mu)G^{\mu\nu} + L'_G(G). & (7.35)
\end{aligned}
$$

These Lagrangians differ only in a term quadratic in the coordinate of the meter:

$$L_F(A, G) - L_A(A, G) = L'_G(G) - L_G(G) = -g^2 G_{\mu\nu}G^{\mu\nu}.$$

This is completely analogous to the way in which the Landau–Peierls and Bohr–Rosenfeld measurement schemes differ. This difference is clearly seen, for example, from the analysis made by DeWitt (1964). The difference arises because, compared with Landau and Peierls, Bohr and Rosenfeld introduced an additional quadratic term in the Lagrangian of the meter in order to 'compensate' for certain errors of the measurement.

Thus, we can interpret the studies of Landau and Peierls (1931) and Bohr and Rosenfeld (1933) in the following way. In the LP scheme a measurement is made of $\partial_\mu A_\nu - \partial_\nu A_\mu$, and this measurement is subject to an absolute limit (7.21)–(7.23); in the BR scheme the tensor $F_{\mu\nu}$ is measured, and there are no absolute limitations on the measurability.

At first glance, one might conclude from this that the BR scheme is better, as has been assumed by most who have discussed this question (see for example DeWitt 1964, Bergmann and Smith 1982, von Borzeszkowski and Treder 1988). However, careful analysis shows that this judgement is not absolute but depends on the point of view or, rather, on the aims that are followed when the electromagnetic field is measured.

We consider first of all the case when the measurement of the electro-magnetic field strength is not the aim in itself but is used to estimate some other field interacting with the electromagnetic field. Let us denote this field by Φ. Measurement of either $\partial_\mu A_\nu - \partial_\nu A_\mu$ or $F_{\mu\nu}$ will yield a certain estimate for Φ. Since the 'curl' of the potential can be measured only with an accuracy not greater than (7.21)–(7.23), restrictions also appear in the estimate of the field Φ. If the field strength $F_{\mu\nu}$ is measured, such restrictions do not arise. Of course, there could be restrictions due to the quantum properties of the field Φ itself. However, as long as the quantum properties of the field Φ can be ignored, the estimate of this field by means of measurement of $F_{\mu\nu}$ can be arbitrarily accurate. Thus, measurement of the field strength (BR scheme) has an advantage (over measurement of the 'curl' of the potential, LP scheme) if the aim of the measurement is to estimate an external classical field (in other words, to make an indirect measurement of this field).

However, a different situation is possible, namely when the measurement of the electromagnetic field strength itself is the aim. In this case, it is important to remember that $\partial_\mu A_\nu - \partial_\nu A_\mu$ describes the state of the electromagnetic field directly: it is a kinematic characteristic of the field. By measuring the 'curl' of the potential, we directly estimate the internal state of the field, and the quantum limit (7.21)–(7.23) shows the extent to which this internal state admits objective estimation.

The tensor $F_{\mu\nu}$ characterizes the internal state of the field indirectly, since it is coupled to the 'curl' of the potential as a dynamical quantity. Therefore, if we measure the field tensor $F_{\mu\nu}$ (the BR scheme) we can estimate an internal state of the field only indirectly: as a result of measurement of $F_{\mu\nu}$ we estimate $\partial_\mu A_\nu - \partial_\nu A_\mu$, and this ultimately gives us information about the state of the field. It follows from the Lagrangian (7.33) that the field tensor and the 'curl' of the potential are connected by

$$\partial_\mu A_\nu - \partial_\nu A_\mu = F_{\mu\nu} + 2gG_{\mu\nu}.$$

Thus, measurement of $F_{\mu\nu}$ enables us to estimate $\partial_\mu A_\nu - \partial_\nu A_\mu$ only up to an accuracy

$$\Delta(\partial_\mu A_\nu - \partial_\nu A_\mu) = \Delta F_{\mu\nu} + 2g\Delta G_{\mu\nu}.$$

To reduce $\Delta F_{\mu\nu}$, it is necessary to increase the coupling constant g, but at the same time the term $2g\Delta G_{\mu\nu}$ increases. The choice of the optimal value of the coupling constant has the consequence that the error in the estimate of the 'curl' of the potential cannot be less than some absolute limit. Thus, the estimate of the internal state of the electromagnetic field by means of the tensor $F_{\mu\nu}$ comes up against quantum limitations, just as in the case when the 'curl' of the potential is measured. The BR scheme has no advantage over the LP scheme in this case.

To make the foregoing discussion clearer, we note that the absence of quantum limits on the measurability of $F_{\mu\nu}$ follows from the Lagrangian (7.33) by virtue of the fact that the error $\Delta F_{\mu\nu}$ can be made arbitrarily small by an increase in the coupling constant g even if the error $\Delta G_{\mu\nu}$ remains finite. As we have just seen, this does not provide an accurate estimate of $\partial_\mu A_\nu - \partial_\nu A_\mu$.

Our final conclusion from this discussion can be formulated as follows:

> A measurement scheme of LP type allows direct estimation of the internal state of an electromagnetic field, but is associated with the absolute quantum limit (7.21)–(7.23). A measurement scheme of BR type permits one to estimate with any degree of accuracy the field tensor $F_{\mu\nu}$ and, through it, the configuration of a classical field interacting with the electromagnetic field. But an estimate of the internal state of the electromagnetic field itself still comes up against quantum limitations. One may suppose that they are described by the same expressions (7.21)–(7.23), though this remains to be proved.

Remark 2 Although in the analysis we have compared our results with those of Landau and Peierls (1931) and Bohr and Rosenfeld (1933), we have here actually studied pointwise measurement of the field in the space-time region Ω, while in the cited papers the problem of measuring the average field in this region was considered. The path-integral approach has been applied to the latter problem by Mensky (1989). The results are the same as those obtained here.

Remark 3 The difference between the variables $\partial_\mu A_\nu - \partial_\nu A_\mu$ and $F_{\mu\nu}$ arising in the context of quantum measurement is quite analogous to the difference between momentum p and velocity (multiplied by a mass) $m\dot{q}$ of a free particle, which has been analysed in detail in section 6.3 of Chapter 6.

Remark 4 Equations (7.21)–(7.23) present restrictions on the measurability of an electromagnetic field following from the quantum properties of this field. Additional restrictions may in principle follow from the discrete character of matter, namely from necessity of making a measurement with the help of particles (say, electrons) having definite characteristics $(m, e, \ldots)$.

8

Time in Quantum Cosmology

In this chapter the path-integral approach to the quantum theory of continuous measurements will be applied to investigate the quantum Universe, taking into account its self-measurement. It will be shown that the problem of time in quantum cosmology can be solved on the basis of such a consideration.

The Wheeler–DeWitt equation is usually exploited to describe the quantum Universe. However, this equation leads to trivial dynamics or, in other words, to no time evolution at all. Time may be obtained from a semiclassical (WKB) analysis of the quantum Universe dynamics. Instead of this we shall show that time emerges in a natural way if self-measurement of the Universe is taken into account.

After some general consideration the amplitude (propagator) describing the dynamics of the quantum Universe will be evaluated explicitly for a simple superspace model. It will be shown that a nontrivial time evolution of the quantum Universe emerges naturally as a result of its self-measurement. The Wheeler–DeWitt dynamics turns out to be valid for small time when the effect of measurement can be neglected and any time dependence disappears.

The chapter is organized in the following way. The problem of time in quantum cosmology is formulated in section 8.1. The way to apply the path-integral theory of continuous measurements in quantum cosmology is sketched in section 8.2 in a general form and is developed in section 8.4 in the framework of a minisuperspace model, introduced in section 8.3. The analysis of the propagator thus obtained is carried out in section 8.5, and a short conclusion is drawn in section 8.6.

8.1 STATEMENT OF THE PROBLEM

Since the work of Hartle and Hawking (1983) on the wavefunction of the Universe there has been great interest in quantum cosmology (see for example Hartle 1988, Calzetta 1989, Unruh 1989, Halliwell 1989, Padmanabhan 1989, Zeh 1989, Grishchuk 1989 and references therein). Work in quantum cosmology is conventionally based on the Wheeler–DeWitt equation, resulting in a trivial dependence of the wavefunction of the Universe on time. One can say that there is no concept of time in quantum cosmology. This leads to difficulties in deriving the time evolution of the quantum Universe and in the transition from quantum to classical cosmology.

Recent work by Zeh (1986, 1988b) and Kiefer (1987) considered self-measurement of the quantum Universe in an explicit way, with the aim of introducing the classical concept of time. Self-measurement can be understood as measuring some of the degrees of freedom of the Universe with the help of the others.

The purpose of this chapter is to treat self-measurement of the quantum Universe in the framework of the phenomenological path-integral quantum theory of continuous measurements (along the lines of the paper by Mensky (1990a)). This allows one to explicitly introduce geometrically defined time into quantum cosmology.

Besides general formulas, the propagator describing the evolution of the quantum Universe under measurement of geometry will be evaluated and analysed in the framework of a simple minisuperspace model, in which geometry is described by a lapse function and a scale factor. This propagator, depending on time in a nontrivial way, generalizes one obtained by Halliwell (1988) and coincides with Halliwell's propagator for sufficiently small time. Only in this limit does the propagator satisfy the Wheeler–DeWitt equation.

The problem of time in quantum cosmology arises from the fact that the Wheeler–DeWitt equation (DeWitt 1967, Wheeler 1968)

$$\hat{H}\Psi = 0 \qquad (8.1)$$

is valid in this theory, leading to the Universe wave Ψ having no dependence on time. This in turn is a consequence of the nonphysical character of the time parameter in gravity. Indeed, any other parameter enumerating the same or some other set of time-slices of the Universe is equally appropriate in classical gravity due to the invariance of the theory under general coordinate transformations. This leads to the Hamiltonian constraint $H(p, q) = 0$ in classical gravity. In quantum cosmology it results in the absence of a picture of time evolution.

To overcome the difficulty it was proposed to use the fact that the Wheeler–DeWitt equation is hyperbolic, with the scale factor a (the size of the Universe) being a time-like coordinate in this equation. The parameter a can therefore serve as an 'intrinsic time'. Another proposal is a probabilistic definition of time in quantum cosmology (Castagnino 1988). However, the problem is still under discussion.

Some time ago Joos (1986) argued that matter effectively measures geometry. Zeh (1986) proposed to utilize a measurement of this type (i.e. the self-measurement of the Universe as a quantum system) to introduce time into quantum cosmology. He used the measurement model based upon a description of the quantum Universe in terms of multipoles as had been suggested by Halliwell and Hawking (1985).

The idea of the self-measurement of the Universe leading to the emergence of time in quantum cosmology was elaborated further by Zeh (1988b) and Kiefer (1987). In these as well as in some subsequent works (see for example Halliwell 1989, Padmanabhan 1989) the phenomenon of decoherentization was investigated, resulting from the self-measurement and leading to a classical character of the parameter a. However, no parameter of time has been introduced except the 'intrinsic time' a . This leads to a tautological character of the concept of the expansion of the Universe. Some other difficulties arise too, for example in the description of returning points of the scale factor, when expansion of the Universe changes into contraction (Halliwell 1989).

We shall show that it is possible to introduce time as a quantitative characteristic of the geometry obtained as a result of self-measurement. Such a time parameter is independent of the scale factor a. In what follows, this program will be realized with the help of the path-integral theory of continuous quantum measurements.

The main idea is that the theory of a quantum system is incomplete without an explicit description of the measurement of this system. A complete picture of the evolution of a quantum system is impossible without explicit consideration of continuous (prolonged in time) measurements of this system (see Chapter 4). The same should be valid for the quantum Universe. Taking the continuous measurement of the Universe into account may lead to a modification of its dynamics such that time dependence will arise naturally.

The only important (in this respect) difference of the Universe from any other quantum system is that no other system can exist outside the Universe. Therefore no measuring device for the Universe can exist outside it. The only kind of measurement possible for the Universe is its self-measurement. This means that some degrees of freedom of the Universe are supposed to play the role of a measuring device with respect to the other degrees of freedom. A model of the self-measurement of the Universe has been considered by Zeh (1986, 1988b) and Kiefer (1987). It is the measurement of the parameters a and φ (the scale factor and scalar field) by higher multipoles of matter and geometry.

Note that self-measurement is also possible in ordinary quantum mechanics if the roles of the measuring device and the measured system are played by different degrees of freedom of the same material system. In fact even the simple and well-known Stern–Gerlach experiment is an example of self-measurement. Indeed, if the projection of the atom's magnetic moment is determined from the deflection of the atom itself in a magnetic field, then the magnetic moment of the atom is a measured system, while its centre of mass is a measuring device. Therefore self-measurement of an atom occurs in the Stern–Gerlach experiment.

We shall consider a continuous measurement (monitoring) of the scale factor a in the framework of the path-integral approach to continuous quantum measurements that does not require explicit description of a measuring device. Instead of this, the measurement is characterized by the information it produces about state of the quantum system (the Universe in our case). Technically the approach reduces to evaluation of the path integral over a restricted set of paths or over all paths but with an appropriate weight functional in path-integral measure.

Calculation in the framework of a simple minisuperspace model shows that the propagator of the quantum Universe acquires, under continuous measurement, a nontrivial time dependence and may even, in certain conditions, satisfy the time-dependent Schrödinger equation. The time-independent Wheeler–DeWitt dynamics turns out to be valid at small time when the effect of measurement is negligible.

8.2 MEASUREMENTS IN QUANTUM COSMOLOGY

The path-integral approach can be applied to describe the measurement of a quantum field. The measurement of curvature in the framework of quantum gravity has been considered in such a way by Mensky (1985a) using a functional integral over configurations of the gravitational field. We shall now apply this approach to quantum cosmology.

An amplitude describing the dynamics of the quantum Universe can be represented as follows:[1]

$$U(^{3}G'', ^{3}G') = \int d\mu[^{3}G] \exp(iS[^{3}G]). \qquad (8.2)$$

Here ^{3}G is a 3-geometry of some time slice (space-like surface) of the Universe, and

$$[^{3}G] = \{^{3}G(\tau) \mid \tau' \leq \tau \leq \tau''\}, \qquad ^{3}G(\tau') = ^{3}G', \quad ^{3}G(\tau'') = ^{3}G'' \qquad (8.3)$$

is the 'path in the space of 3-geometries', i.e. a (3+1)-foliation of the 4-geometry of the Universe between two time slices. $S[^{3}G]$ is a gravitational action corresponding to the 4-geometry expressed by $[^{3}G]$ and $d\mu[^{3}G]$ is a gauge-invariant measure on the space of all 4-geometries. Integration should be carried out over all 4-geometries with the boundary conditions

---

[1]Natural units are used hereafter, in which c, $\hbar$ and the gravitational constant κ are equal to 1.

$^3G'$, $^3G''$. The formula (8.2) is written for pure gravity, but it can be generalized in a staightforward way for gravity interacting with matter. However, we shall not need this generalization here.

The amplitude (propagator) (8.2) describes the dynamics of the quantum Universe in the absence of measurements. It satisfies the Wheeler–DeWitt equation (8.1) where the Hamiltonian $\hat{H}$ is obtained from the action $S[^3G]$. Therefore the dependence of the propagator (8.2) on the parameters τ', τ'' is trivial, and this is a consequence of their arbitrariness (nonphysical character). Suppose, however, that continuous measurement of geometry is carried out during the interval $[\tau', \tau'']$. Then an amplitude should be modified by introducing a weight functional into it:

$$U_{[^3\mathcal{G}]}(^3G'',{}^3G') = \int d[^3G]\, w_{[^3\mathcal{G}]}[^3G]\, \exp(iS[^3G]). \qquad (8.4)$$

Here $[^3\mathcal{G}]$ denotes the 4-geometry describing the result of measurement and the functional $w_{[^3\mathcal{G}]}$ expresses the information contained in this result, about the actual 4-geometry (that is about a path in the space of 3-geometries along which the quantum Universe propagates). This means that $w_{[^3\mathcal{G}]}[^3G]$ is small or zero if the geometry $[^3G]$ is far from $[^3\mathcal{G}]$ and is equal (or close) to unity if these geometries are close to each other.

The idea we consider now is that the physical time between the instants marked by the parameters τ', τ'' can be evaluated from the geometry $[^3\mathcal{G}]$, and thus the propagator (8.4) in fact depends on this physical time in a nontrivial way. This leads to a conventional picture of the time evolution of the Universe. The previous form of propagator (8.2) and the Wheeler–DeWitt regime of trivial time dependence must be valid in the situation when the measurement does not modify the dynamics significantly, that is the functional $w_{[^3\mathcal{G}]}$ is close to unity for those 4-geometries which contribute to the integral (8.2) significantly.

8.3 MINISUPERSPACE WITHOUT MEASUREMENT

Halliwell (1988) evaluated an amplitude (8.2) for the minisuperspace model resulting from a metric of the form

$$ds^2 = -\frac{N^2(\tau)}{q(\tau)}\, d\tau^2 + q(\tau)\, d\Omega_3^2. \qquad (8.5)$$

Then the gravitational field is represented by the only parameter q, which is equal to the square of the scale factor,

$$q = a^2.$$

Here τ is an arbitrary parameter enumerating time slices of the Universe, and the transformation of this parameter should be accompanied by the corresponding transformation of the lapse function N:

$$\begin{aligned}
d\tau &\rightarrow & d\hat{\tau} = w(\tau)\,d\tau, \\
N(\tau) &\rightarrow & \hat{N}(\hat{\tau}) = w^{-1}(\tau)N(\tau).
\end{aligned} \tag{8.6}$$

Time t corresponding to the geometry (8.5) and invariant under the transformation (8.6) can be defined by the equation $dt = N(\tau)\,d\tau$. Note that the 'physical' time equal to a space-time interval between corresponding points of the slices of the Universe is $dt_{\mathrm{phys}} = q^{-1}\,dt = q^{-1}N\,d\tau$. We shall use this parameter at the very end of our consideration.

The Hamiltonian following from the metric (8.5) and Einstein action is

$$H = \frac{1}{2}(-4p^2 + \lambda q - 1) \tag{8.7}$$

where λ is a cosmological constant. Classical trajectories $q^{\mathrm{class}}(\tau)$ defined by this Hamiltonian are parabolas satisfying the Hamiltonian equations

$$\dot{q} = N\{q, H\}, \qquad \dot{p} = N\{p, H\}$$

leading in the present case to the equation

$$\ddot{q} = 2\lambda N^2. \tag{8.8}$$

In addition, the classical trajectory should satisfy the constraint

$$H = 0.$$

Taking the expression (8.7) for a Hamiltonian in (8.2) and constructing a gauge-invariant measure with the help of ghosts, then taking the gauge $dN/d\tau = 0$ and integrating over ghosts, Halliwell (1988) obtained for the propagator (8.2) the following integral:

$$U(q'', q') = (\tau'' - \tau')\int d[\Pi]\,d[N]\,d[p]\,d[q]\exp\left(i\int_{\tau'}^{\tau''} d\tau\,(p\dot{q} - NH + \Pi\dot{N})\right)$$

where integration over Π should be carried out as over the coordinate with homogeneous boundary conditions while N should be integrated as a corresponding momentum. After integration over Π one obtains

$$U(q'', q') = (\tau'' - \tau')\int d[N]\,\delta[\dot{N}]\,d[p]\,d[q]\exp\left(i\int_{\tau'}^{\tau''} d\tau\,(p\dot{q} - NH)\right).$$

Integration over $[p]$ can be also carried out explicitly to give

$$U(q'',q') \;=\; (\tau''-\tau')\int \mathrm{d}[N]\,\delta[\dot{N}]\,\mathrm{d}[q]$$

$$\times\; \exp\left[-\frac{\mathrm{i}}{8}\int_{\tau'}^{\tau''}\mathrm{d}\tau\left(\frac{\dot{q}^2}{N}+4N\lambda q-4N\right)\right]. \qquad (8.9)$$

Below we shall modify the integral (8.9) to take the measurement into account, but now let us consider its unmodified form, as in the paper of Halliwell (1988). Integration over $[q]$ can be carried out explicitly and integration over $[N]$ can be reduced (with the aid of $\delta[\dot{N}]$) to an integral over a single numerical value N (this possibility exists because the gauge $\dot{N}=0$ has been accepted for the calculation). It turns out that this number occurs only in the combination $(\tau''-\tau')N$ in the integrand. Denoting this combination by

$$T=(\tau''-\tau')N$$

one obtains

$$U(q'',q') = \int \mathrm{d}T\, G_T(q'',q') \qquad (8.10)$$

where

$$G_T(q'',q') = \frac{1}{\sqrt{8\pi\mathrm{i}T}}\exp\left[-\mathrm{i}\left(\frac{(q''-q')^2}{8T}-\frac{T}{2}-\frac{\lambda T(q'+q'')}{4}-\frac{\lambda^2 T^3}{24}\right)\right].$$

It can be easily verified that this propagator is a solution to the Wheeler–DeWitt equation

$$\frac{1}{2}\left(4\frac{\partial^2}{\partial q''^2}+\lambda q''-1\right)U(q'',q')=0$$

if integration over T goes from $-\infty$ to ∞, and is a Green's function of this equation,

$$\frac{1}{2}\left(4\frac{\partial^2}{\partial q''^2}+\lambda q''-1\right)U(q'',q')=-\mathrm{i}\delta(q''-q'),$$

if this integration goes from 0 to ∞.

8.4 MINISUPERSPACE UNDER MEASUREMENT

Let us now modify the integral (8.9), taking the self-measurement of the Universe into account. The physical models for self-measurement were discussed briefly in section 8.1 and have been investigated in detail by Zeh

(1986, 1988b) and Kiefer (1987). Here we shall make use of the path-integral approach, allowing one to treat continuous measurements with no explicit model of a measuring device. One of the advantages of this approach is that it enables one to take into account the influence of the measurement without knowing the actual scheme of measurement. (We should know only what information the measurement gives.) This is important in the present case because we do not know precisely what measurements are made in the Universe (if any).

In the framework of the present minisuperspace model the geometry is described by the metric (8.5) depending on the functions $[q]$ and $[N]$. Therefore the self-measurement means that some functions $[\kappa]$, $[\nu]$ are found as estimates of the corresponding functions $[q]$ and $[N]$. The measurement guarantees that the actual functions $[q]$ and $[N]$ differ from their estimates by no more than some given entities (the errors of the corresponding measurements). This can be expressed with the help of the following functionals:

$$u_{[\kappa]}[q] = \exp\left(-\frac{1}{\rho^2}\int_{\tau'}^{\tau''} d\tau\, N(\tau)(q-\kappa)^2\right)$$

$$v_{[\nu]}[N] = \exp\left(-\frac{1}{\sigma^2}\int_{\tau'}^{\tau''} d\tau\, |N-\nu|\right). \qquad (8.11)$$

Of course, the pairs of functions $([\kappa],[\nu])$, $([\kappa'],[\nu'])$ correspond to the same measurement output if one of them can be obtained from the other one by reparametrization (8.6) (where ν should be substituted for N). This is taken into account by choosing the functionals (8.11) to be invariant under the reparametrization (8.6).

With the functionals (8.11) we have for a propagator

$$U_{[\kappa,\nu]}(q'',q')$$

$$= (\tau''-\tau')\int d[N]\,\delta[\dot N]\,d[q]\exp\left[-\frac{i}{2}\int_{\tau'}^{\tau''} d\tau\left(\frac{\dot q^2}{4N}+N\lambda q - N\right)\right]$$

$$\times \exp\left[-\int_{\tau'}^{\tau''} d\tau\left(\frac{|N-\nu|}{\sigma^2}+N\frac{(q-\kappa)^2}{\rho^2}\right)\right].$$

Integration over $[N]$ can be reduced, with the help of $\delta[\dot N]$, to integration over the number N. To integrate over $[q]$, let us represent this function as a sum of two functions, $q = q_0 + z$, where $z(\tau') = z(\tau'') = 0$, and q_0 satisfies the following equation and boundary conditions:[2]

$$\ddot q_0 = 0, \qquad q_0(\tau') = q', \qquad q_0(\tau'') = q''.$$

---

[2] Using this equation for q_0 turns out to be technically more convenient than using the classical equation (8.8) though the latter is also possible.

Then

$$U_{[\kappa,\nu]}(q'',q')$$

$$= (\tau''-\tau')\int dN \exp\left[-\int_{\tau'}^{\tau''} d\tau \left(\frac{|N-\nu|}{\sigma^2}+\frac{i}{8N}\dot{q}_0^2\right)+\frac{i}{2}N(\tau''-\tau')\right]$$

$$\times \int_0^0 d[z]\exp\left[-i\int_{\tau'}^{\tau''} d\tau \left(\frac{\dot{z}^2}{8N}+\frac{1}{2}\lambda Nz - iN\frac{(z-\kappa+q_0)^2}{\rho^2}\right)\right].$$

Integration over $[z]$ can be performed by transforming to a spectral representation (see section 3.3 of Chapter 3), that is to integration over the variables $z_1, z_2, \ldots$, defined by

$$z(\tau) = \sum_{k=1}^{\infty} z_k \sin\Omega_k(\tau-\tau'), \quad z_k = \frac{2}{\tau''-\tau'}\int_{\tau'}^{\tau''} d\tau\, z(\tau)\sin\Omega_k(\tau-\tau')$$

$$(8.12)$$

with

$$\Omega_k = \frac{k\pi}{\tau''-\tau'}.$$

This gives finally

$$U_{[\kappa,\nu]}(q'',q') = \int dT\, W(T/\rho)\,\Sigma_T[\nu]\,R_T[\kappa]\,G_T(q'',q'). \qquad (8.13)$$

Here the function

$$W^2(x) = \frac{2(1+i)x}{\sin[2(1+i)x]}$$

and the functionals

$$\Sigma_T[\nu] = \exp\left(-\frac{1}{\sigma^2}\int_{\tau'}^{\tau''} d\tau \left|\nu(\tau)-\frac{T}{\tau''-\tau'}\right|\right)$$

$$R_T[\kappa] = \exp\left(-\frac{T}{2\rho^2}\sum_{k=1}^{\infty}\frac{(\kappa_k-q_{Tk})^2}{1-8i\left(\frac{T}{\pi k\rho}\right)^2}\right)$$

have been introduced.

The latter functional is expressed in terms of the decompositions κ_k, q_{Tk} (in a series of the type of (8.12) for the function $\kappa(\tau)$ and the solution $q_T(\tau)$ of the classical equation of motion (8.8) with the lapse function $N = T/(\tau''-\tau')$:

$$\ddot{q}_T = \frac{2\lambda T^2}{(\tau''-\tau')^2}, \qquad q_T(\tau') = q', \qquad q_T(\tau'') = q''.$$

Note that the function $q_T(\tau)$ is not in general a classical trajectory because it does not satisfy the constraint $H = 0$. For this constraint to be satisfied T should be equal to T_+ or T_- where

$$T_\pm = \lambda^{-1/2}(Q'^2 + Q''^2 \pm 2Q'Q''), \quad Q''^2 = q'' - \lambda^{-1}, \quad Q'^2 = q' - \lambda^{-1}.$$

8.5 ANALYSIS OF THE RESULTS

The propagator (8.13) describing the minisuperspace dynamics under the self-measurement differs from (8.10) by three factors in the integrand, W, Σ_T and R_T. Let us analyse these factors and their role. Our aim is to find conditions providing that these factors are not small so that the measurement amplitude (8.13) is close to maximum.

The function $W(x)$ is approximately equal to unity while $|x| \lesssim 1/2$. The absolute value of $W(x)$ decreases monotonically, becomes less than $1/2$ when $|x| > 2$ and then decays very quickly (exponentially). These properties will be important for further analysis.

The functional Σ_T characterizes a distribution of different possible functions $[\nu]$ describing (together with the functions $[\kappa]$) the results of the measurement of geometry. It can be easily seen that $\Sigma_T[\nu]$ can be close to unity for some value of T only if the function $[\nu]$ is almost constant. Otherwise Σ_T is much less and the corresponding measurement output has a small probability. This is a consequence of the accepted gauge condition $\dot N = 0$.

For further analysis let us take $\nu = \text{const}$. Then

$$\Sigma_T[\nu] = \Sigma(T - t) = \exp\left(-\frac{|T - t|}{\sigma^2}\right) \tag{8.14}$$

where the notation

$$t = (\tau'' - \tau')\nu$$

is introduced. Since the value ν is an estimation of the lapse function, t is evidently an estimation of the actual (gauge-invariant) duration of the interval $[\tau', \tau'']$. Thus physically significant time emerges in our analysis, and (8.13) takes the form

$$U_{[\kappa],t}(q'', q') = \int dT\, W(T/\rho)\, \Sigma(T - t)\, R_T[\kappa]\, G_T(q'', q'). \tag{8.15}$$

The functional R_T in (8.15) characterizes the distribution of different functions $[\kappa]$ estimating the squared scale factor $[q]$. To obtain a general notion about the character of this functional, let us neglect the imaginary term in the denominator of its exponent. This term describes quantum effects in the measurement of $[q]$ and should generally be taken into account. It is negligible for

$$t \leq 1, \qquad \sigma^2 \leq \rho.$$

However, the character of the functional $R_T[\kappa]$ and its role can be estimated without the quantum term even for other values of the parameters. Neglecting this term and taking $T = t$ (because of (8.14)), one has

$$R_t[\kappa] \simeq R_t^0[\kappa] = \exp\left(-\frac{t}{2\rho^2}\sum_{k=1}^{\infty}(\kappa_k - q_{tk})^2\right).$$

By the Parseval formula, this is equal to

$$R_t^0[\kappa] = \exp\left(-\frac{t}{\rho^2(\tau'' - \tau')}\int_{\tau'}^{\tau''} d\tau\,(\kappa - q_t)^2\right)$$

with $q_t = q_t(\tau)$.

This means that the result of measurement $[\kappa]$ with large probability should be close to the parabola

$$q_t(\tau) = \lambda t^2\left(\frac{\tau - \tau'}{\tau'' - \tau'}\right)^2 + [(q'' - q') - \lambda t^2]\frac{\tau - \tau'}{\tau'' - \tau'} + q'$$

(the propagator becomes exponentially small otherwise). Considering $R_t[\kappa]$ for a fixed function $[\kappa]$ but for different q', q'', we see that these points turn out to be localized about the corresponding end points of the measured scale factor $[\kappa]$, as one could expect. More precisely, they are localized about the corresponding end points $\bar{q}'$, $\bar{q}''$ of the parabola $\bar{q}_t(\tau)$ which maximizes $R_t[\kappa]$ when substituting for $q_t(\tau)$.

Let us now suppose that q', q'' are sufficiently close to the corresponding points $\bar{q}'$, $\bar{q}''$. Then $R_T[\kappa]$ in (8.15) may be taken equal to unity. Let us investigate the dependence of the propagator on q', q'' and on t due to the other factors. The propagator in this case takes the form

$$U_{[\kappa],t}(q'', q') = \int dT\, W(T/\rho)\,\Sigma(T - t)\,G_T(q'', q'). \qquad (8.16)$$

For a fixed value of t, the function $\Sigma(T - t)$, defined by (8.14), is large for T close to t and exponentially small when T differs from t by more than σ^2. For small σ the function $\Sigma(T - t)$ in (8.16) can be approximately substituted by a δ-function:

$$\Sigma(T - t) \simeq 2\sigma^2\delta(T - t).$$

This is possible if all other factors in the integrand of (8.16) can be considered to be constant when T differs from t by an amount of the order of

σ^2. This is the case if

$$\sigma \ll 1, \quad \lambda\sigma^2 q' \ll 8, \quad \lambda\sigma^2 q'' \ll 8, \quad t \ll \rho, \quad \frac{\sigma|q''-q'|}{2\sqrt{2}} \ll t \ll \frac{2\sqrt{2}}{\lambda\sigma}. \tag{8.17}$$

Then the propagator takes the form

$$U_t(q'', q') = 2\sigma^2 G_t(q'', q'). \tag{8.18}$$

It can be easily verified that such a propagator satisfies the equation

$$\frac{1}{2}\left(4\frac{\partial^2}{\partial q''^2} + \lambda q'' - 1\right) U_t(q'', q') = i\frac{\partial}{\partial t} U_t(q'', q'). \tag{8.19}$$

It is seen from (8.7) that this equation can be rewritten as

$$i\frac{\partial}{\partial t} U_t(q'', q') = \hat{H} U_t(q'', q'). \tag{8.20}$$

Therefore the dependence of the wavefunction of the Universe on the time t appearing as a result of the measurement is described by the time-dependent Schrödinger equation, provided the inequalities (8.17) are fulfilled. Yet it is possible that the conditions (8.17) for this regime of propagation are not valid in reality.

Let us now consider the opposite limit of large σ. Suppose the following inequalities are fulfilled:

$$\rho^2 \gg 1, \quad \sigma^2 \gg 1, \quad t \ll \sigma^2, \quad q' \ll \lambda^{-1}, \quad q'' \ll \lambda^{-1}. \tag{8.21}$$

Then the functions $\Sigma(T-t)$ and $W(T/\rho)$ are equal to unity for all values of T that are important for integration (i.e. in the region where oscillations of the factor $\exp(\frac{1}{2}iT)$ are negligible). Therefore, under these conditions the propagator (8.16) becomes equal to Halliwell's propagator (8.10) satisfying the Wheeler–DeWitt equation.

It can be seen from (8.21) that Halliwell's propagator (8.10) is a sufficiently good approximation only for sufficiently small time ($t \ll \sigma^2$) and not too accurate self-measurement ($\rho \gg 1$, $\sigma \gg 1$). If the time t is larger than the characteristic threshold σ^2, the influence of the self-measurement should not be neglected, and the more general form of the propagator, (8.16) or (8.15), should be used. However, even for small time, $t \ll \sigma^2$, the Wheeler–DeWitt equation is not applicable if the self-measurement is too accurate, $\rho \lesssim 1$ or $\sigma \lesssim 1$. We do not know what is real accuracy, and this question cannot be answered in the framework of the present phenomenological approach but requires a physical model of self-measurement of the type considered by Zeh (1986, 1988b) and Kiefer (1987).

The parameter $t = (\tau''-\tau')\nu$ used up to now to express the duration of a measurement interval is not a length of a space-time interval. It can be seen from the form of the metric (8.5) that physical or rather geometrical time (proper time), coinciding with the space-time interval, can be introduced as

$$t_{phys} \simeq \frac{t}{a}, \qquad a^2 \simeq \kappa.$$

It is seen from equation (8.14) that the error of measurement of this time is

$$\Delta t_{phys} \simeq \frac{\sigma^2}{a}.$$

Analogously, the equation (8.15) together with (8.14) shows that the error of measurement of the scale factor a is

$$\Delta a \simeq \frac{\rho}{2at^{1/2}}.$$

With the help of these physically more transparent parameters the conditions (8.21) for validity of Halliwell's propagator can be rewritten as follows:

$$t_{phys} \ll \Delta t_{phys}, \quad 4a^3 t_{phys} \Delta a^2 \gg L^6, \quad a\Delta t_{phys} \gg L^2, \quad \lambda a^2 \ll 1. \quad (8.22)$$

Here ordinary units are used (but $c = 1$) and

$$L = \sqrt{\frac{G\hbar}{c^3}} \simeq 10^{-33} \text{ cm}$$

is the Planck length. The conditions (8.17) for the propagator to have the simple form (8.18), satisfying the Schrödinger equation, are then

$$a\Delta t_{phys} \ll L^2, \qquad \lambda a^3 \Delta t_{phys} \ll L^2$$

$$\frac{a^{3/2}\Delta t_{phys}^{1/2}}{3L} \ll t_{phys} \ll \frac{3L}{\lambda a^{3/2}\Delta t_{phys}^{1/2}}. \qquad (8.23)$$

In conclusion, let us make two remarks.

Remark 1 The parabolas $q_t(\tau)$ arising in the course of our consideration satisfy the equation of motion (8.8) following from the Hamiltonian (8.7) but do not generally satisfy the constraint $H = 0$. This constraint can be traced in the form of the quantum propagator (8.13), namely in the fact that oscillations (in T) of the function $G_T(q'', q')$ are minimum near the values $T = T_{\pm}$. This yields that the propagator $U_t(q'', q')$ is smaller for t other than $T_{\pm}$. However, for sufficiently small values of σ^2 oscillations of $G_T(q'', q')$ become less important. For example, they are unimportant in the Schrödinger regime (8.17). Therefore sufficiently precise measurement of the geometry of the Universe in the time-like direction suppresses the classical constraint $H = 0$. The Universe may in fact develop according to a non-classical scenario in this case. An example is the choice of $\lambda = 0$ in the minisuperspace model under consideration, when no real classical trajectory exists but evolution under measurement is possible.

Remark 2 We did not consider above the problems arising due to the fact that the variable $q = a^2$ is restricted by the inequality $q \geq 0$, thus requiring certain boundary conditions at $q = 0$. These conditions can be taken into account, for example, by the method of images (see Halliwell 1988 for a general discussion of this and other methods). If the $q = 0$ boundary conditions are taken into account, the form of the propagator (8.13) will alter for small q', q''. However, the conclusion drawn above about the emergence of time will of course be valid, as well as all the main points of the above analysis.

8.6 CONCLUDING REMARKS

We have considered here the possible influence of self-measurement on the dynamics of the quantum Universe. The influence could be caused by reduction of the wavefunction of the Universe in the process of continuous measurement. Reduction of this sort can be taken into account naturally in the framework of the path-integral approach to quantum measurement. For the sake of clarity the actual calculations have been carried out for a simple minisuperspace model, considered (without measurement) in the work of Halliwell (1988). It was shown that the measurement of geometry described by the scale factor and lapse function leads to the natural emergence of time as a measure of an interval between two space-like slices of the Universe. Therefore the time evolution of the quantum Universe makes sense in quantum cosmology provided its self-measurement is correctly taken into account.

This is one of the illustrations (maybe the most exciting one) of the general statement that a complete description of a quantum system should include its continuous measurement (see also section 4.3 of Chapter 4).

Remark 3 The measurement does not influence the dynamics of the system if it is performed in the classical regime. This has been discussed in Chapters 4 and 5 (see also Chapters 9 and 11). Interesting arguments in connection with this can be found in the paper by Gell-Mann and Hartle (1990) where the concept 'decohering sets of histories' is used instead of the term 'classical regime of measurement'. In fact they consider no measurement at all. However, measurement in the classical regime leads to no state reduction. Therefore the description of the system in terms of decohering sets of paths and the consideration of measurement in the classical regime are essentially identical situations.

The above consideration shows that the quantum Universe may be described by Halliwell's propagator (8.10) (i.e. by the Wheeler–DeWitt equation (8.1)) only for sufficiently small time when the influence of the self-measurement is negligible (8.21). For longer time the more general propagator (8.15) should be used, which takes the self-measurement of the Universe into account. This regime is characterized by a nontrivial dependence on time. Under some conditions (8.17) (perhaps too severe to be valid in reality) this time-dependent propagator takes a simple form (8.18) satisfying the time-dependent Schrödinger equation (8.20).

An important question arises of course about the actual values of the parameters σ and ρ, or more generally about the real regime of the self-measurement of the quantum Universe. This question cannot be answered in the framework of the present phenomenological approach. Instead, it requires an investigation of physical models of self-measurement. An example of such a model can be found in the papers of Zeh (1986, 1988b) and Kiefer (1987).

9

The Action Uncertainty Principle

The path-integral approach to the quantum theory of continuous measurements has been developed in the preceding chapters. According to this approach the measurement amplitude determining the probabilities of different outputs of the measurement can be evaluated in the form of a restricted path integral (a path integral 'in finite limits'). With the help of the measurement amplitude, the maximum deviation of the measurement outputs from that predicted by classical theory can be easily determined. The aim of this chapter is to express this deviation, i.e. the variance of outputs, in the simpler and more transparent form of a specific type of uncertainty principle (called the action uncertainty principle, AUP).

The simplest (but weakest) form of AUP is $\delta S \gtrsim \hbar$ where S is the action functional. A stronger form (with wider application) of AUP (for ideal measurements performed in the quantum regime) is $|\int_{t'}^{t''} \frac{\delta S[q]}{\delta q(t)} \Delta q(t) \, dt| \simeq \hbar$, where the paths $[q]$ and $[\Delta q]$ stand respectively for the measurement output and the measurement error. This can also be presented in symbolic form as $\delta(\text{Equation})\Delta(\text{Path}) \simeq \hbar$. This means that the deviation of the observed (measured) motion from that obeying the classical equation of motion is inversely proportional to the uncertainty in a path (the latter uncertainty resulting from the measurement error).

The consequence of AUP is that improving the measurement precision beyond the threshold of the quantum regime leads to less information resulting from the measurement. Examples are given below of applications of AUP to continuous measurements in quantum mechanics and quantum gravity. In the latter case Rosenfeld's restriction on the measurability of curvature is derived from AUP almost trivially.

9.1 UNCERTAINTIES IN CONTINUOUS MEASUREMENTS

The formalism of Feynman path integrals has been applied to continuous quantum measurements in the preceding chapters and proved to be natural and efficient. An amplitude U_α for each output α of a continuous measurement can be presented in the form of a path integral in such a

138

way that $P_\alpha = |U_\alpha|^2$ is a probability density for different outputs of the measurement. The most probable is of course the output α_{class} predicted by classical theory. However, deviation of the outputs α from the classical one α_{class} is possible. The variance $\delta\alpha$ of probable outputs around α_{class} is one of the most important predictions of the theory. Of course the variance $\delta\alpha$ of the outputs depends on the error $\Delta\alpha$ of the measurement.

It turns out that (partly simplifying the situation) there is a quantum threshold $\Delta\alpha_{quant}$ such that for rough measurements ($\Delta\alpha \gg \Delta\alpha_{quant}$) a classical regime of measurement occurs while for fine measurements ($\Delta\alpha \ll \Delta\alpha_{quant}$) the quantum regime arises. In the classical regime $\delta\alpha \simeq \Delta\alpha$, i.e. the variance of the measurement outputs is determined completely by the measurement errors. However, in the quantum regime $\delta\alpha \simeq \Delta\alpha_{quant}^2/\Delta\alpha$. In this case the variance of outputs is a consequence of quantum fluctuations (or of the unavoidable back reaction of a measuring device on the measured quantum system). This 'quantum measurement noise' increases as the measurement error $\Delta\alpha$ decreases. This rather paradoxical phenomenon is the reason why there is no sense in decreasing the measurement error $\Delta\alpha$ beyond the quantum threshold $\Delta\alpha_{quant}$. The optimal regime of measurement is on the frontier between the classical and quantum regimes, when $\Delta\alpha \simeq \Delta\alpha_{quant}$. In this regime the variance of outputs $\delta\alpha$ is minimum and equal to $\delta\alpha_{opt} \simeq \Delta\alpha_{quant}$. This is an absolute limit for the given type of measurement.

Of course the scheme of the preceding paragraph is only a rough qualitative outline of a more complicated real situation that varies from one measurement to another. Two main corrections required to this rough scheme are the following: (1) the quantum threshold $\Delta\alpha_{quant}$ and the measurement regimes are to be defined separately for each frequency range of the process or for the given form of a force acting on the system under consideration; (2) there is a type of measurement, called quantum nondemolition (QND) measurement, which has no quantum threshold at all. QND measurements are in fact always performed in the classical regime (see Chapter 6 and section 9.6 of the present chapter).

Any continuous measurement can be completely analysed if the measurement amplitudes U_α are evaluated for all measurement outputs α. However, this requires the calculation of a path integral. The aim of this chapter is to give another, simpler form of such an analysis following the papers by Mensky (1991, 1992a, b). It turns out that the variance of the measurement outputs $\delta\alpha$ satisfies a rather simple inequality $\delta S \gtrsim \hbar$ including the action functional S. This inequality resembles the known uncertainty principle in many respects. This is why it could be called 'the action uncertainty principle' (AUP).

A stronger form of AUP (with a much greater area of application) can be derived for a linear system with the help of a linear decomposition (for example in frequency components). It can be written as an equality

$$\left| \int_{t'}^{t''} \frac{\delta S[q]}{\delta q(t)} \Delta q(t) \, dt \right| \simeq \hbar$$

valid for an ideal measurement performed in the quantum regime.

9.2 CLASSICAL REGIME OF MEASUREMENT

Let us briefly repeat the argument leading to the path-integral approach to quantum continuous measurements. The amplitude (called a *propagator*) $U(q'', q')$ for a system (one may think for example about a particle) to move from the point q' to the point q'' can be expressed in the form of the sum (or rather integral) of the amplitudes $U[q]$ corresponding to all possible paths

$$[q] = \{q(t) | t' \le t \le t''\}$$

connecting the points q' and q'':

$$U(q'', q') = \int d[q] \, U[q]. \tag{9.1}$$

However, this expression for the propagator is valid only if there is no possibility of finding out which path is actually followed as a channel for the transition from q' to q''. Let us suppose, however, that a *continuous measurement* is performed simultaneously with this transition, giving some information about the path of transition. Such information can be expressed by some set of paths I_α. If the measurement gives the result α then the transition occurs via one of the paths $[q]$ belonging to the set I_α. Then the amplitude for the transition from q' to q'' can be expressed as an integral over paths belonging to I_α:

$$U_\alpha(q'', q') = \int_{I_\alpha} U[q] \, d[q]. \tag{9.2}$$

If α is fixed, the amplitude $U_\alpha = U_\alpha(q'', q')$ can be considered as the propagator of the system undergoing the continuous measurement (with the given output). If q', q'' are fixed, the same amplitude can be thought of as the probability amplitude for the continuous measurement to give the result α. Taking a square modulus $P_\alpha = |U_\alpha|^2$ of the amplitude, one can obtain the probability density for different results of the continuous measurement.

The Heisenberg uncertainty principle $\Delta q \Delta p \gtrsim \hbar$ is convenient for expressing the features of instantaneous quantum measurements, but it is inconvenient for continuous measurements. Indeed, if the measurement is, for example, monitoring the coordinate q up to some error Δq, then it gives some information about momentum too, but it is difficult to express this information quantitatively and apply the uncertainty principle to derive restrictions on the efficiency of continuous measurements.

Therefore it is natural to try to obtain restrictions of the type of the uncertainty principle but for continuous measurements. In all cases these restrictions can be derived from path integrals of the type of equation (9.2). Let us now analyse some general features of such a calculation and try to formulate the uncertainty principle for continuous measurements.

The amplitude $U[q]$ corresponding to a separate path is, according to Feynman,

$$U[q] = e^{\frac{i}{\hbar}S[q]}$$

where $S[q]$ is the action functional for the system under consideration. The path integral (9.2) can therefore be rewritten as follows:

$$U_\alpha = \int_\alpha e^{\frac{i}{\hbar}S[q]}\, d[q] \qquad (9.3)$$

where we have omitted the arguments q', q'' and identified the result of the measurement α with the corresponding set of paths I_α. In the rest of this chapter α will be identified with I_α.

The value U_α can be interpreted as an amplitude for the measurement to give the result α. This means that only those measurement results are probable for which U_α is large in absolute value. We shall try to estimate this value by analysing the behaviour of the action functional $S[q]$.

In what follows we shall suggest that all sets α have the same 'width' and differ only in their forms. Then the value of U_α depends only on how rapidly $S[q]$ varies when $[q]$ stays in the limits of α.

Remark 1 The idea is that the 'measure' of the set α, calculated with the help of

$$d[q] = \prod_t dq(t)$$

is the same for all α so that U_α depends only on oscillations of the factor $\exp\left(\frac{i}{\hbar}S[q]\right)$ when $[q] \in \alpha$. The statement about the 'measure' $d[q]$ can be given meaning with the aid of the limiting process. A more precise formulation is that the measure $d[q]$ is invariant under the group G of time-localized shifts (see Chapter 10),

$$g = \{c(t)|\, t' \le t \le t''\}, \qquad g[q] = \{q(t) + c(t)|\, t' \le t \le t''\},$$

and all sets α can be presented in the form $\alpha = g\alpha_0$ with different $g \in G$ and the same α_0.

It is known that only paths close in a sense to the classical trajectory contribute to the unrestricted Feynman integral

$$U(q'', q') = \int e^{\frac{i}{\hbar} S[q]} \, d[q]. \qquad (9.4)$$

This is because the action functional $S[q]$ is extremal on the classical trajectory $[q_{\text{class}}]$:

$$\frac{\delta S[q]}{\delta q(t)}\bigg|_{[q]=[q_{\text{class}}]} = 0.$$

Indeed, $S[q]$ changes slowly in the vicinity of $[q_{\text{class}}]$ and all partial amplitudes $U[q] = \exp\left(\frac{i}{\hbar}S[q]\right)$ have almost equal phases in this vicinity, which leads to 'constructive interference'.

The constructive interference occurs in the limits of the set

$$I_{\text{class}} = \{[q] \mid |S[q] - S[q_{\text{class}}]| \lesssim \hbar\}$$

and changes into 'destructive interference' outside this set when oscillation becomes rapid. Therefore, the main contribution to the Feynman path integral is given by the set of paths I_{class}:

$$U(q'', q') \simeq \int_{I_{\text{class}}} d[q] \, e^{\frac{i}{\hbar} S[q]}.$$

It is evident (by analogous argument) that the measurement amplitude (9.3) has its maximum value for $\alpha = \alpha_{\text{class}}$ where the set α_{class} contains the classical trajectory as its middle path.

Remark 2 The latter affirmation can be made strict with the help of the above mentioned group G of time-localized shifts. In terms of this group $\alpha_{\text{class}} = V[q_{\text{class}}]$ where the subset $V \subset G$ is symmetric in respect with unity, i.e. contains g^{-1} together with any g.

Now the classical regime of measurement can be defined by the requirement that

$$\alpha_{\text{class}} \supset I_{\text{class}}. \qquad (9.5)$$

In this case the integral $U_{\alpha_{\text{class}}}$ is practically equal to the complete Feynman integral U. It is evident that U_α remains the same for $\alpha \supset I_{\text{class}}$. However, U_α decays when an intersection $\alpha \cap I_{\text{class}}$ decreases, becoming almost null for $\alpha \cap I_{\text{class}} = \emptyset$. The reason is that the amplitude $U[q]$ oscillates rapidly in this case in all regions of the set α.

From the argument of the preceding paragraph one can see that the probable outputs of the measurement performed in the classical regime (9.5) may be characterized by the condition

$$[q_{\text{class}}] \in \alpha. \qquad (9.6)$$

This condition takes into account the sets α including I_{class} and those having a significant intersection with I_{class}. The only sets excluded by the condition are those having small or null intersection with I_{class}.

The condition (9.6) completely corresponds to the conclusions of classical measurement theory, as could be expected in the case of the classical regime of measurement. In fact, all the predictions of classical measurement theory are valid for continuous measurements performed in the classical regime.

It is important for the derivation of the action uncertainty principle that in the case of the classical regime of measurement any measurement output α emerging with rather high probability contains the set I_{class} or most of it (in this context we of course interpret the measurement output α as a corresponding set of paths). It then follows from the definition of I_{class} that the variance of the action S in the limits of the set α (corresponding to a probable measurement output) is not less than the quantum of action:

$$\Delta_\alpha S \gtrsim \hbar.$$

This will be used in the next section.

9.3 QUANTUM REGIME AND AUP

Consider now the quantum regime of measurements when no set α includes I_{class}. The most probable in this case is the output corresponding to the set α_{class} lying in the middle of I_{class}. However, another α is probable too provided that the variation of the action in the limits of this set is small enough, $\Delta_\alpha S \lesssim \hbar$. Therefore for some of the measurement outputs α having high probability, the variation of S is at the boundary of the area characterized by the preceding inequality, $\Delta_\alpha S \simeq \hbar$. Taking all probable (emerging with high probability) sets α and estimating the variation δS of S for all of them we obtain for this variation $\delta S \gtrsim \hbar$.

This conclusion is drawn for the quantum regime of measurement. However, in the classical regime $\Delta_\alpha S \gtrsim \hbar$ for each probable measurement output α. We conclude therefore that for the measurement outputs arising with comparatively high probability the variance δS of S in any regime of measurement satisfies the following inequality:

$$\delta S \gtrsim \hbar. \tag{9.7}$$

This is the simplest form of the *action uncertainty principle*, AUP. However, it turns out to be too weak (though we shall apply it in section 9.7 to the measurement of a quantum gravitational field). A stronger inequality expressing AUP can be derived for the case of a linear system (it is valid also for a nonlinear one provided that the sets α are narrow enough).

In the case of a linear system different components of some linear decomposition of the motion can be analysed separately. For a given measurement some of these components may turn out to be measured in the classical regime while other components are measured (by the same measurement) in the quantum regime.

Let us for definiteness use decomposition into a series of frequency components. A continuous measurement gives information about each frequency component of a path $[q]$. Therefore the measurement can be thought of as a number of measurements, each of them measuring a definite frequency component q_Ω. A typical situation is when some frequency components q_Ω are measured in the classical regime ($\alpha_\Omega \supset I_\Omega^{\text{class}}$) while the others are measured in the quantum regime.

Let the system under consideration be linear, i.e. the Lagrangian $L(q, \dot{q})$ and the action $S[q]$ be quadratic. Then, expanding a path $[q]$ in frequency components with frequencies Ω and amplitudes q_Ω, we have an expansion for the action too:

$$S[q] = \sum_\Omega S_\Omega(q_\Omega).$$

The measure for integrating over paths factorizes as follows (see section 3.3 of Chapter 3):

$$d[q] = \prod_\Omega dq_\Omega.$$

As a result, the integral (9.4) is factorized in the following way:

$$U = \prod_\Omega U_\Omega, \qquad U_\Omega = \int dq_\Omega \, e^{\frac{i}{\hbar} S_\Omega(q_\Omega)}.$$

Suppose that the restriction of a path $[q]$ inside the path set α can also be expressed in terms of frequency components (otherwise another linear expansion may be taken),

$$q_\Omega \in \Delta_\Omega(a_\Omega) = [a_\Omega - \Delta a_\Omega, a_\Omega + \Delta a_\Omega].$$

Here the set of intervals Δ_Ω for frequency components determines the measurement output α.

The restricted integral (9.2) then takes the following factorized form:

$$U_\alpha = \prod_\Omega U_\Omega(a_\Omega), \qquad U_\Omega(a_\Omega) = \int_{\Delta_\Omega(a_\Omega)} dq_\Omega \, e^{\frac{i}{\hbar} S_\Omega(q_\Omega)}. \qquad (9.8)$$

In this formula the measurement is interpreted as the measurement of frequency components. The output of the measurement of the frequency component q_Ω is denoted by a_Ω, and the error of this measurement is denoted by Δa_Ω. An interval $\Delta_\Omega(a_\Omega)$ is equivalent to the pair $(a_\Omega, \Delta a_\Omega)$.

The most probable are classical outputs a_Ω^{class} forming the expansion for the classical path q_{class}. This means that $U_\Omega(a_\Omega)$ has its maximum absolute value for $a_\Omega = a_\Omega^{\text{class}}$. This is because the partial action $S_\Omega(q_\Omega)$ has an extremum at $q_\Omega = a_\Omega^{\text{class}}$ and oscillation of the integrand in (9.8) is minimum in this point.

Denote

$$I_\Omega^{\text{class}} = \{q_\Omega \mid |S_\Omega(q_\Omega) - S_\Omega(a_\Omega^{\text{class}})| \lesssim \hbar\}.$$

Then the classical regime of Ω-measurement is defined by

$$\Delta_\Omega(a_\Omega^{\text{class}}) \supset I_\Omega^{\text{class}}$$

and the quantum regime emerges otherwise. In the classical regime, outputs a_Ω such that $a_\Omega^{\text{class}} \in \Delta_\Omega(a_\Omega)$ are probable. The latter condition may also be written as $a_\Omega^{\text{class}} \in \alpha$ since $\Delta_\Omega(a_\Omega)$ is a frequency component of α.

Consider now the quantum regime of measurement of the Ω-component, when $\Delta_\Omega(a_\Omega^{\text{class}})$ is narrower than I_Ω^{class}. In this case the variation ΔS_Ω of S_Ω in the limits of $\Delta_\Omega(a_\Omega^{\text{class}})$ is less than $\hbar$. Therefore $\exp\left(\frac{i}{\hbar} S_\Omega(q_\Omega)\right)$ is almost constant and the amplitude of the most probable output a_Ω^{class} is approximately equal to the measure of the integration range:

$$|U_\Omega(a_\Omega^{\text{class}})| \simeq \Delta a_\Omega. \tag{9.9}$$

Going over to another measurement output a_Ω that differs from a_Ω^{class}, one sees that the partial measurement amplitude U_Ω remains the same while the oscillation of the exponent in the limits of Δ_Ω remains small. The condition for this is that

$$|\Delta S_\Omega| \lesssim \hbar$$

for

$$\Delta S_\Omega = S_\Omega(q_\Omega) - S_\Omega(a_\Omega)$$

with any q_Ω in the limits of Δ_Ω.

The latter condition should be summed up over all Ω to give the following condition for the measurement output α to be probable:

$$\left| \sum \Delta S_\Omega \right| \lesssim \hbar.$$

Here summation is taken only over those Ω for which $a_\Omega \neq a_\Omega^{\text{class}}$.

The expression in the sum can be given the form

$$\left. \frac{\partial S_\Omega(q_\Omega)}{\partial q_\Omega} \right|_{q_\Omega = a_\Omega} \Delta q_\Omega$$

with $\Delta q_\Omega = q_\Omega - a_\Omega$. This has the advantage that this expression is automatically equal to zero for $a_\Omega = a_\Omega^{\text{class}}$, so that the preceding sum may be taken over all Ω:

$$\left| \sum_{\Omega} \frac{\partial S_{\Omega}(q_{\Omega})}{\partial q_{\Omega}} \right|_{q_{\Omega}=a_{\Omega}} \Delta q_{\Omega} \lesssim \hbar. \tag{9.10}$$

The deviations Δq_{Ω} here can be arbitrary but not driving $a_{\Omega} + \Delta q_{\Omega}$ from $\Delta_{\Omega}(a_{\Omega})$. If the inequality is valid for any possible choice of deviations then the output α is probable.

9.4 AUP AND EQUATIONS OF MOTION

Returning from the frequency components to the paths themselves, one can rewrite the condition (9.10) in the form of an integral:

$$\left| \int_{t'}^{t''} \frac{\delta S[q]}{\delta q(t)} \Delta q(t) \, dt \right| \lesssim \hbar. \tag{9.11}$$

Here $[q]$ is the middle path of the set of paths α, and $[\Delta q]$ is a deviation of $[q]$ which does not drive it from α, so that $[q+\Delta q] \in \alpha$ too. If the inequality (9.11) is fulfilled for a given set α and for any deflection $[\Delta q]$ not driving from α, then the corresponding measurement output α emerges with high probability.

Since equation (9.11) is a condition for the output α to be probable, there are some probable outputs α for which the left-hand side of equation (9.11) is of the order of $\hbar$. Thus

$$\left| \int_{t'}^{t''} \frac{\delta S[q]}{\delta q(t)} \Delta q(t) \, dt \right| \simeq \hbar \tag{9.12}$$

is valid for probable outputs of the measurement performed in the quantum regime. Here $q(t)$ is the middle path of the corresponding output α and $\Delta q(t)$ is a typical deflection from this path, leading to the maximum value of the left-hand side of (9.11). The inequality (9.11) and the equality (9.12) present another form of the uncertainty principle for continuous measurements (the *action uncertainty principle*, AUP). How this principle can be applied to actual systems and measurements will be seen from examples given in subsequent sections.

The entity

$$\frac{\delta S[q]}{\delta q(t)} = \frac{\partial L(q, \dot{q})}{\partial q(t)} - \frac{d}{dt} \frac{\partial L(q, \dot{q})}{\partial q(t)} \tag{9.13}$$

is nothing but the left-hand side of the classical equation of motion:

$$\frac{\delta S[q]}{\delta q(t)} = 0.$$

This is why the equality (9.12) can be symbolically rewritten as follows:

$$\delta(\text{Equation})\Delta(\text{Path}) \simeq \hbar.$$

This means that the deviation from the classical picture must be small for rough measurements (when $\Delta(\text{Path})$ is large) but it can be large for fine measurements (with small $\Delta(\text{Path})$). Fine measurements performed in the quantum regime give rise to non-classical behaviour of the system (we mean behaviour that can be estimated by the measurement output).

If the left-hand side (9.13) of a classical equation is not equal to zero, then one could say that a fictitious force $\delta F(t)$ has arisen equal to this nonzero value. The equality (9.12) can therefore be rewritten as follows:

$$\left| \int_{t'}^{t''} \Delta q(t)\delta F(t)\,\mathrm{d}t \right| \simeq \hbar. \tag{9.14}$$

Here the left-hand side of AUP is expressed as an integral of the left-hand side $\Delta p(t)\Delta q(t)$ of an ordinary uncertainty principle with

$$\Delta p(t) = \delta F(t)\,\mathrm{d}t$$

defined by a fictitious force $\delta F(t)$ arising as a result of the measurement noise.

The qualitative difference of AUP from the Heisenberg uncertainty principle is that the former deals with the variance of outputs of a single measurement, not of two measurements (of complementary variables) as does the latter. This of course is a direct consequence of this single measurement being continuous, and thus containing information about both complementary observables. For example, monitoring of position gives some information about both position and momentum of the system during the measurement interval.

9.5 EXAMPLES FROM QUANTUM MECHANICS

We shall consider here two simple examples of the application of AUP in quantum mechanics: position monitoring of a free particle and of a harmonic oscillator. Only the case of a one-dimensional particle and oscillator will be considered. However, generalization to the multidimensional case is straightforward.

9.5.1 Monitoring of the Position of a Free Particle

Consider a (one-dimensional) free particle having a classical equation of motion of the form

$$m\ddot{q} = F. \tag{9.15}$$

The fictitious force emerging in equation (9.14) and describing quantum measurement noise coincides with the left-hand side of this equation, $\delta F = m\ddot{q}$. For monitoring the coordinate with the error Δa we have $|\Delta q| = \Delta a$ for a typical deflection of a path inside the corridor. Therefore equation (9.14) in this case takes the form

$$\Delta a \int_{t'}^{t''} |\delta F(t)|\, dt \simeq \hbar. \qquad (9.16)$$

To make use of this estimation, one should accept some hypothesis about the form of the force. Suppose that the real force acting on the particle is constant, $F = \text{const}$. Then a reliable conclusion about this force on the basis of the measurement output is possible only if this force is greater than the fictitious force *of the same form* arising as measurement noise. It follows from (9.16) that constant fictitious force

$$\delta F = \delta F_0 = \text{const}$$

can emerge if its value $\delta F_0 \lesssim \delta F_{\text{quant}}$ where

$$\delta F_{\text{quant}} = \frac{\hbar}{T\Delta a} \qquad (9.17)$$

where we have denoted $T = t'' - t'$. This is the magnitude of the quantum measurement noise in the estimation of a constant force by monitoring of the position of a free particle.

To obtain the classical measurement noise, one should use the classical equation of motion (9.15) and the solution to this equation for a constant force:

$$q(t) = q(t') + \frac{1}{2}\frac{F}{m}(t - t')^2.$$

The maximum deflection of the coordinate under the influence of the force during the time period T is equal to

$$q(t'') - q(t') = \frac{1}{2}\frac{F}{m}T^2.$$

Monitoring of the coordinate with an error Δa therefore allows (according to classical theory) the force F to be estimated with an error of the order of

$$\delta F_{\text{class}} = \frac{m\Delta a}{T^2}. \qquad (9.18)$$

The total error in the estimation of a constant force is

$$\delta F_{\text{total}} = \delta F_{\text{class}} + \delta F_{\text{quant}} = \frac{m\Delta a}{T^2} + \frac{\hbar}{T\Delta a}.$$

We see that this error is purely classical (equal to (9.18)) for $\Delta a \gg \Delta a_{\mathrm{opt}}$ where

$$\Delta a_{\mathrm{opt}}^2 = \frac{\hbar T}{m} \qquad (9.19)$$

and is purely quantum (equal to (9.17)) for $\Delta a \ll \Delta a_{\mathrm{opt}}$. In the first case we have the classical regime of measurement, and in the second case it is the quantum regime.

If the measurement error Δa decreases from initially very big values, then the error of the force estimation $\delta F_{\mathrm{total}}$ decreases until Δa achieves the quantum threshold Δa_{opt} and then increases again. The optimal regime of measurement corresponds to the measurement error Δa of the order of Δa_{opt}. The error of the force estimation is equal in this case (to the order of magnitude) to

$$\delta F_{\mathrm{opt}} = \sqrt{\frac{\hbar m}{T^3}}. \qquad (9.20)$$

The error in the estimation of position is in this case equal to

$$\delta q_{\mathrm{opt}} \simeq \Delta a_{\mathrm{opt}} = \sqrt{\frac{\hbar T}{m}}. \qquad (9.21)$$

The formulas (9.21), (9.20) give the so-called *standard quantum limit* (SQL) for the measurability of the position of a free particle and a constant force acting on this particle. The SQL has been derived by Braginsky (1967)[1] (see also Braginsky and Vorontsov 1974, Mensky 1979a, 1979b, Caves *et al* 1980, Caves 1985). In the framework of the path-integral approach this limit was derived by Mensky (1979a).

Remark 3 The present approach guarantees that the SQL is absolute, i.e. it cannot be overcome by the choice of the measurement scheme, provided that the measurement gives the information specified (path $[q]$ with some accuracy Δa in our case). This conclusion is valid, however, only if a rather narrow definition of measurement is accepted, corresponding to the operation of restriction of a path integral. According to this definition the measured observable (a coordinate in our case) remains in the limits indicated by the measurement output during the whole measurement period. Some authors accept a weaker definition, suggesting that measurement gives some information about the state of the system before the measurement but that the state during the measurement (and in some cases after it) does not correspond to this information. If this definition is accepted, the formulas derived here are not valid.

---

[1] In this original paper an expression for the SQL was obtained for the case of an oscillator, $\delta F = \sqrt{\hbar \omega m}/T$ (see section 5.3.3 of Chapter 5). The free particle SQL can be considered to be a special case of this formula. It follows if the characteristic time T is substituted for ω^{-1}.

Consider the same system (a free particle) but with another hypothesis about the form of the force. Let the force acting on the particle have the form of a short pulse of duration $\tau \ll T$. Then we are interested in a fictitious force δF of the same form:

$$\delta F(t) = \begin{cases} \delta F_0 & \text{for } t_1 \le t \le t_1 + \tau \\ 0 & \text{otherwise} \end{cases}$$

for some t_1 between t' and $t'' - \tau$. Substituting this for δF in (9.16), we obtain the estimation of δF_0 equal to

$$\delta F_{\text{quant}} = \frac{\hbar}{\tau \Delta a}. \tag{9.22}$$

This is quantum measurement noise in the case under consideration.

The classical measurement noise can be found by considering the motion of a classical free particle under the influence of a short pulse of a force. If the magnitude of a force in the pulse is F then the linear momentum it transfers to the particle is $F\tau$, and the displacement of the particle under the influence of this pulse is equal to

$$q(t'') - q(t_1) \simeq \frac{F\tau}{m} T$$

(we accepted here that $t'' - t_1$ is of the order of T). Taking into account that this displacement can be observed only with the error Δa, we have for the minimum magnitude of the force which can be reliably measured

$$\delta F_{\text{class}} = \frac{m \Delta a}{\tau T}. \tag{9.23}$$

The total measurement noise is the sum of (9.23) and (9.22). The optimal regime corresponds to the choice $\Delta a = \Delta a_{\text{opt}}$ when the two terms in this sum are equal. In the optimal regime one has

$$\delta F_{\text{opt}} \simeq \frac{1}{\tau} \sqrt{\frac{m\hbar}{T}}, \qquad \delta q_{\text{opt}} \simeq \Delta a_{\text{opt}} = \sqrt{\frac{\hbar T}{m}}. \tag{9.24}$$

These formulas give an absolute restriction on measurement of the given type and are nothing but a standard quantum limit for this measurement (Braginsky and Vorontsov 1974, Mensky 1979a).

9.5.2 Monitoring of the Position of an Oscillator

Now let the system under investigation be a harmonic oscillator of mass m and frequency ω. Then

$$\delta F = m(\ddot{q} + \omega^2 q).$$

If the continuous measurement is monitoring of the coordinate q with the error Δa, then the typical deflection of a path in equation (9.14) corresponds to $|\Delta q(t)| \leq \Delta a$. The left-hand side of this inequality is maximum if one chooses $\Delta q(t) = \Delta a$ for $\delta F(t) \geq 0$ and $\Delta q(t) = -\Delta a$ for $\delta F(t) \leq 0$. This brings equation (9.14) into the form (9.16).

The fictitious force $\delta F(t)$ satisfying this inequality can arise as measurement noise. The actual force acting on the oscillator is observable if it is greater than the fictitious force arising as measurement noise.

To estimate the minimum observable force one has to know (or suggest) its form. For example, let the force be harmonic with frequency Ω. The fictitious harmonic force

$$\delta F(t) = \delta F_0 \cos(\Omega t + \phi)$$

can arise, according to equation (9.16), if $\delta F_0 \simeq \hbar/T\Delta a$ where $T = t'' - t'$ is the time of measurement. This gives an estimation for the minimum observable force in the quantum regime of measurement: $\delta F_{\text{quant}} = \hbar/T\Delta a$. The corresponding estimation for the classical regime can be obtained in the framework of classical measurement theory, starting from the equation $\delta F = m(\ddot{q} + \omega^2 q)$. This gives $\delta F_{\text{class}} = m|\Omega^2 - \omega^2|\Delta a$.

Taking both classical and quantum measurement noise into account, one has

$$\delta F_{\text{total}} = \delta F_{\text{class}} + \delta F_{\text{quant}} = m|\Omega^2 - \omega^2|\Delta a + \frac{\hbar}{T\Delta a}.$$

In the classical regime of measurement (when Δa is large enough) the second term of this sum is negligible, while in the quantum regime (small Δa), in contrast, the second term dominates. The optimal regime of measurement corresponds to $\Delta a = \Delta a_{\text{opt}}$, where

$$\Delta a_{\text{opt}}^2 = \frac{\hbar}{Tm|\Omega^2 - \omega^2|}.$$

In the optimal regime the minimum is achieved of an observable force equal to (up to the order of magnitude) δF_{opt}, where

$$\delta F_{\text{opt}}^2 = \frac{m\hbar}{T}|\Omega^2 - \omega^2|.$$

This is in accordance with the results of sections 5.2 and 5.3 of Chapter 5.

9.6 QUANTUM NONDEMOLITION MEASUREMENTS

The example considered above revealed a typical situation with a continuous measurement when there is an optimal precision of measurement on the border between the classical and quantum regimes of measurement. Both rougher and finer measurements are disadvantageous as compared with the optimal measurement, the former because of classical and the latter because of quantum measurement noise. The variance of the measurement outputs corresponding to the optimal level of measurement puts an absolute quantum limit on the given type of measurement. This corresponds to the concept of a *standard quantum limit* (SQL) in quantum measurement theory.

To overcome the SQL, so-called *quantum nondemolition* (QND) measurements have been proposed by Braginsky *et al* (1977). It has been shown by Golubtsova and Mensky (1989) that for this type of measurement there is in fact no quantum regime (see Chapter 6). This means that no absolute restriction on observable force arises in this case, contrary to the example of section 9.5. Let us consider the simplest type of QND measurement with the help of the action uncertainty principle.

Consider the same harmonic oscillator as in section 9.5 but the other continuous measurement. Choose monitoring of the quadrature component X of the oscillator, one of the two quadrature components

$$X = q\cos\omega t - \frac{p}{m\omega}\sin\omega t$$
$$Y = q\sin\omega t + \frac{p}{m\omega}\cos\omega t,$$

both of which are QND observables, complementary to each other.

Since the observable X contains both q and p, its treatment requires a phase-space (or Hamiltonian) representation of path integrals. The measurement amplitude corresponding to the measurement output α has in this representation, instead of equation (9.3), the form

$$U_\alpha = \int_\alpha \exp\left(\frac{i}{\hbar}\int(p\dot{q} - H(p,q))\,dt\right)d[q].$$

Here α denotes the measurement output and simultaneously the corresponding set of paths $[p,q]$ in phase space. H is the Hamiltonian of the system:

$$H(p,q) = \frac{p^2}{2m} + \frac{m\omega^2 q^2}{2}$$

in the case of a harmonic oscillator.

The action uncertainty principle can be presented in terms of the Hamiltonian as follows:

$$\left|\int_{t'}^{t''}\left[\left(\dot{q} - \frac{\partial H}{\partial p}\right)\Delta p - \left(\dot{p} + \frac{\partial H}{\partial q}\right)\Delta q\right]dt\right| \simeq \hbar. \qquad (9.25)$$

The analogy of this formula with equation (9.12) is evident. However, the Hamiltonian (canonical) equations of motion now appear (instead of

Lagrangian equations) together with the deviations $[\Delta q]$, $[\Delta p]$ of the two constituents $[q]$, $[p]$ of the phase-space path $[p, q]$.

If one expresses $\Delta p, \Delta q$ in terms of $\Delta X, \Delta Y$,

$$\Delta p = \frac{\partial p}{\partial X}\Delta X + \frac{\partial p}{\partial Y}\Delta Y$$

$$\Delta q = \frac{\partial q}{\partial X}\Delta X + \frac{\partial q}{\partial Y}\Delta Y$$

the left-hand side of equation (9.25), with the help of expressions for X and Y, can be put in the form

$$m\omega \int_{t'}^{t''} \left[\dot{Y}(t)\Delta X(t) - \dot{X}(t)\Delta Y(t)\right]\, dt. \tag{9.26}$$

If only the observable X is measured, then ΔX is finite and $\Delta Y = \infty$ (measurement with infinite error means no measurement at all). This is why we should take

$$\dot{X} = 0 \tag{9.27}$$

so that the second term in equation (9.26) disappears.[2] After this argument equation (9.25) may be written in the form

$$m\omega \left| \int_{t'}^{t''} \dot{Y}(t)\Delta X(t)\, dt \right| \simeq \hbar. \tag{9.28}$$

According to the classical equation of motion the quadrature component Y (as well as X) is an integral of motion, so that

$$\dot{Y} = 0 \quad \text{or} \quad Y = \text{const}$$

should be true for a classical oscillator. The equation (9.28) then shows that the deviation from the conservation law for Y is inversely proportional to the error of the measurement of X. For monitoring X with the error Δx one has

$$\delta Y = \frac{\hbar}{m\omega\Delta x}.$$

What about the other quadrature component X? Its measurement gives, according to equation (9.27), the classical picture of conservation of this variable.

It is important that measurement of X has an influence on the dynamics of the complementary variable Y but not on X itself. As a result,

---

[2] This is quite analogous to the procedure that has been used to justify omitting some terms on the left-hand side of AUP in the general consideration of a linear system. We may think that the observable Y is measured in the classical regime and therefore can be omitted just like those terms in the sum (9.10) which correspond to classically measured frequency components.

the variance of the measurement outputs for X lies in the limits of the measurement error Δx, which may be arbitrary small. No absolute restriction on the observability of this variable arises. In fact its measurement is essentially classical. This is characteristic of QND measurements.

9.7 AUP IN QUANTUM GRAVITY

The path-integral approach has already been applied to measurements of quantum fields: an electromagnetic field (see Chapter 7) and a gravitational field (Mensky 1985a). Measurements of this type can be called continual because they are prolonged in time and protracted in space. Calculations of amplitudes for such measurements ought to be based upon the integral over field configurations instead of the integral over trajectories (though the term 'path integral' is often accepted in this context too). Now we shall show that application of the action uncertainty principle makes evaluation of the variance for outputs of some continual measurements very easy. The measurement of the mean curvature in the framework of quantum gravity will be taken as an example.

Take the Einstein expression for the action of a gravitational field in the form

$$S = \frac{c^3}{16\pi G} \int_\Omega R(-g)^{1/2}\, \mathrm{d}^4 x.$$

Accepting the four-volume of the space-time region Ω to be L_0^4 and denoting the scalar curvature averaged over this region by $\overline{R}$, one has

$$S = \frac{c^3}{16\pi G} \overline{R}\, L_0^4.$$

Then, taking AUP in the form $\delta S \gtrsim \hbar$ (as in equation (9.7)), one obtains

$$\delta \overline{R}\, L_0^4 \gtrsim 16\pi L_{\mathrm{Pl}}^2 \tag{9.29}$$

with

$$L_{\mathrm{Pl}} = \sqrt{\frac{G\hbar}{c^3}} \simeq 10^{-33}\ \mathrm{cm}$$

being the Planck length.

The inequality (9.29) is nothing but Rosenfeld's uncertainty relation for measurement of the mean curvature. It was derived by Rosenfeld (1966) (see also von Borzeszkowski 1982) with the help of a concrete model of measurement and by Mensky (1985a) in the framework of the path-integral approach. We see that the same relation can be derived much more easily with the help of AUP.

9.8 CONCLUDING REMARKS

A simple inequality has been derived above as a condition for the continuous measurement outputs to be probable. This inequality expresses a sort of an uncertainty principle for continuous measurements and can be called the action uncertainty principle (AUP). Its simple form is equation (9.7). Another, and in a sense more transparent, form is equations (9.11), (9.12).

The principal distinctions of this new uncertainty principle from the Heisenberg uncertainty relation should be emphasized. The latter restricts the variance of outputs of two different measurements (for example of position and momentum). But the output of a continuous measurement contains information about both position and momentum (or any other pair of conjugate observables). Therefore some restriction should arise on the variance of outputs of this measurement alone. This restriction is expressed by AUP. It turns out that (in the quantum regime of measurement) the more accurate the continuous measurement, the more its outputs can differ from those predicted by classical theory, as can be seen from equation (9.12).

The qualitative consequences of this principle arising in a typical situation can be formulated as follows:

1. For a rough measurement (when the error of the measurement is large and the sets α wide) only those measurement outputs are probable which are predicted by classical theory. This is a classical regime of measurement.

2. For a fine measurement (when the error is small and the sets α narrow) even those measurement outputs that are far from classical predictions are probable. The more precise the measurement, the wider is the range of probable measurement outputs (this may be called quantum measurement noise). This is a quantum regime of measurement.

3. The optimal regime lies between the classical and quantum regimes of measurement. Rougher measurements are inefficient because of classical measurement errors. Finer measurements are inefficient because of the quantum measurement noise.

This leads to an absolute restriction on the measurability of the corresponding observable. This restriction cannot be overcome by the choice of the measuring device for the same type of measurement. One can, however, choose the type of measurement or the observable to be measured in such a way that an absolute restriction does not arise. In this case the observable and the measurement are called quantum nondemolition (QND). For any such measurement there is in fact no quantum regime of measurement.

Therefore no quantum threshold arises in this case and there is no absolute restriction on observable force. QND measurements can be considered in the framework of the path-integral approach provided the phase-space (Hamiltonian) representation of path integrals is used.

10

Group-Theoretical Structure of Quantum Continuous Measurements

In this chapter quantum continuous measurements will be investigated from the point of view of their group-theoretical structure. It turns out that there are two such kinds of structure in each measurement. The 'transverse' group transforms alternative measurement outputs into each other and thus describes the homogeneity of the space of outputs. This is important for choosing the measure of integration over all measurement outputs.

The other group-theoretical structure can be called 'longitudinal'. It acts in the direction of time and describes the evolution of the quantum system subject to continuous measurement. This structure is described by a semigroup instead of a group, so that not all elements have an inverse. This reflects the fact that time evolution has a definite direction and cannot (in nonrelativistic theory) go backward.

In addition to a short exposition of each of these structures their unification in a single semigroup will be considered. The resulting group-theoretical scheme generalizes the scheme connected with the evolution of a nonrelativistic particle in an external field.

The subject of this chapter needs further development. In fact the reader will find here only some hints about possible links between two large areas: the theory of quantum continuous measurements and quantum theory based upon the formalism of the group of paths.

10.1 EVOLUTION UNDER MEASUREMENT

The group-theoretical approach in quantum theory, especially in elementary particle theory, is widespread and has given important results in quantum field theory in recent decades (Wigner 1939, Newton and Wigner 1949, Wightman 1962, see also Mensky 1976, 1983a). This chapter is devoted to an analysis of group-theoretical structure arising in the description of quantum continuous measurements, along the lines of the paper by Mensky (1990b).

There are two sources of a group-theoretical structure. One is that evolution develops continuously in time. We shall show that evolution under continuous measurement can be described by a semigroup, each element of which characterizes not only the time interval but also the measurement output obtained during this interval. This semigroup will be called the longitudinal one.

Another source of group-theoretical structure is the equivalence of different measurement outputs. This equivalence can be described by the transverse group transforming alternative outputs into each other. The terms 'longitudinal' and 'transverse' arose from the most typical example of continuous measurement, namely the coordinate monitoring expressed by the diagram of coordinate versus time. The time direction is then longitudinal and the coordinate direction is transverse.

We know from section 4.3 of Chapter 4 that a quantum continuous measurement and the evolution of a quantum system subject to such a measurement can be described with the help of a path integral. To do this, integration in the path-integral expression for a propagator should be restricted by a weight functional. It will be convenient for us now to change notation, introducing the measurement output α not as a subscript but as an argument of the weight functional:

$$U_\alpha(q'', q') = \int_{q'}^{q''} d[q] \exp\left(\frac{i}{\hbar} S[q]\right) w(\alpha, [q]). \tag{10.1}$$

Here $w(\alpha, [q])$ is a functional of $[q]$ close to unity for paths $[q]$ compatible with the information contained in the output α and close to zero otherwise. The measurement interval is supposed to be $[t', t'']$.

Instead of a propagator as a two-point function we can use the corresponding operator (with this two-point function as a kernel):

$$U(\alpha) = \int d[q] \exp\left(\frac{i}{\hbar} S[q]\right) w(\alpha, [q]) V[q]. \tag{10.2}$$

Here integration is performed over paths with all possible end points and $V[q]$ denotes an operator of translation along the path $[q]$:

$$(V[q]\psi)(q'') = \begin{cases} \psi(q') & \text{if } q(t') = q', \ q(t'') = q'' \\ 0 & \text{if } q'' \neq q(t''). \end{cases} \tag{10.3}$$

If the system under consideration is in the state $|\psi'\rangle$ at the initial instant t', then it will be finally in the state

$$|\psi''(\alpha)\rangle = U(\alpha)|\psi'\rangle \tag{10.4}$$

depending of course on the measurement result α. For the mixed initial state described by the density matrix ρ' the final state is

$$\rho''(\alpha) = U(\alpha)\rho' U^\dagger(\alpha). \tag{10.5}$$

This evolution law is valid if the measurement result is known. If it is not known (non-selective measurement) or if *a priori* calculation of the evolution is to be done then summation should be performed over all alternative measurement results (see section 4.3 of Chapter 4):

$$\rho'' = \int d\mu(\alpha)\, U(\alpha)\rho' U^\dagger(\alpha). \tag{10.6}$$

10.2 TRANSVERSE GROUP FOR ALTERNATIVES

Though some alternative outputs are more probable than others, this is the result of a calculation of measurement amplitude. *A priori* (before calculation of an amplitude) all alternative measurement outputs are to be equivalent. This means that some group G acts transitively on the set of alternative measurement results, and the measure $d\mu(\alpha)$ is invariant under this action. Let us call this group the transverse group and denote the action of the transverse group on alternatives and paths as follows:

$$g : \alpha \to g\alpha, \qquad g : [q] \to g[q].$$

Let us suppose that the measurement functional is invariant in the following sense:

$$w(g\alpha, g[q]) = w(\alpha, [q]). \tag{10.7}$$

This means that the value of $w(\alpha, [q])$ depends only on the 'distance' between the path $[q]$ and the path $[q]_\alpha$ characterizing the output α (for example $[q]_\alpha$ may be the 'middle path' of the corridor expressing the output of the coordinate monitoring). Simultaneous and coordinated change of $[q]_\alpha$ and $[q]$ does not change the above-mentioned distance and therefore the value of $w(\alpha, [q])$. This is what the formula (10.7) means.

In fact the functional $w(\alpha, \cdot)$ completely determines the measurement output α. Therefore, given the action of the transverse group on paths, its action on alternative measurement outputs can be derived from equation (10.7):

$$w(g\alpha, \cdot) = w(\alpha, \cdot) \cdot g^{-1}. \tag{10.8}$$

The action of the transverse group G on alternatives (measurement outputs) is transitive. This is why an arbitrary alternative α can be obtained from the previously given 'standard' alternative α_0 with the help of the corresponding group element, $\alpha = g\alpha_0$. Integration over all alternatives can therefore be reduced to integration over the group G with some measure dg:

$$d\mu(\alpha) = d\mu(g\alpha_0) = dg.$$

It is natural to require that this measure be invariant:

$$d(g_1 g) = dg.$$

If the transverse group is also transitive on the set of all paths, then any path $[q]$ can be obtained from the previously given 'standard' path $[q_0]$ by the action of an appropriate group element, $[q] = g'[q_0]$. Integration over paths can also, under certain conditions, be reduced to integration over the group:

$$d[q] = d(g'[q_0]) = dg'.$$

Denoting

$$w(\alpha_0, g'[q_0]) = w(g')$$

one can reduce the set of all functionals $w(\alpha, \cdot)$ describing the measurement to a single function on the group:

$$w(\alpha, [q]) = w(g\alpha_0, g'[q_0]) = w(\alpha_0, g^{-1}g'[q_0]) = w(g^{-1}g').$$

The operator (10.2) can then be put into the purely group-theoretical form

$$U(\alpha) = U(g\alpha_0) = \int dg' \exp\left(\frac{i}{\hbar}S(g')\right) w(g^{-1}g')V(g')$$

where the following notation is introduced:

$$S(g') = S(g'[q_0]), \qquad V(g') = V(g'[q_0]).$$

Let us consider as an example (actually quite general) of the transverse group the following group of curves (figure 10.1):

$$g = \{c(t)| \, t' \leq t \leq t''\}.$$

Here each $c(t)$ is a shift in the configuration space that the coordinate $q(t)$ belongs to. Then the action of the group on paths is as follows:

$$g[q] = \{q(t) + c(t)| \, t' \leq t \leq t''\}. \tag{10.9}$$

Suppose that the continuous measurement consists in the measurement of a mean coordinate (averaged over the given time interval):

$$\bar{q} = \frac{1}{t'' - t'} \int_{t'}^{t''} q(t)\,dt.$$

Then the measurement result is expressed by the number a (or a vector in the multidimensional case) interpreted as an estimation of the mean coordinate. This type of measurement can be described by the functional

$$w(a, [q]) = W(a - \bar{q}) \tag{10.10}$$

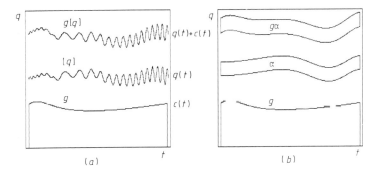

Figure 10.1: The action of a transverse group (a) on paths [q] and (b) on measurement outputs α (for coordinate monitoring as the measurement).

where W is a function expressing properties of the measuring device (its characteristic). It follows then from (10.9) and (10.7) that the action of the transverse group on alternative measurement results consists in their shift:

$$ga = a - \overline{c}, \qquad \overline{c} = \frac{1}{t'' - t'} \int_{t'}^{t''} c(t)\, dt.$$

Another example is the measurement of the spectral components q_n of a path $[q]$ defined as follows:

$$q(t) = \sum_{n=1}^{\infty} q_n \sin \Omega_n (t - t'), \qquad \Omega_n = \frac{\pi n}{t'' - t'}.$$

Then the measurement result is described by the set of numbers $(a_1, a_2, \ldots)$ and

$$(ga)_n = a_n - c_n$$

where c_n are spectral components of the function $c(t)$:

$$c_n = \frac{2}{t'' - t'} \int_{t'}^{t''} c(t) \sin \Omega_n (t - t')\, dt.$$

The functional describing the measurement of the spectral components q_n may have the form

$$w(a_1, a_2, \ldots; [q]) = \prod_{n=1}^{\infty} W_n(q_n - a_n) = \exp\left(\sum_{n=1}^{\infty} B_n(q_n - a_n) \right). \quad (10.11)$$

Consider finally one more example of continuous measurement, namely continuous monitoring of the coordinate. The result of this measurement can be expressed by a curve $[a] = \{a(t)\,|\,t' \leq t \leq t''\}$ in the coordinate (configuration) space. The corresponding functional can be taken to be

$$w([a],[q]) = W\left(\int_{t'}^{t''} (q(t) - a(t))^2 \, dt\right) \qquad (10.12)$$

or even in a more general form

$$w([a],[q]) = W\left(\int_{t'}^{t''} f(q - a) \, dt\right).$$

The action of the transverse group can be shown in the present case to be a shift:

$$g[a] = [a - c] = \{a(t) - c(t)\,|\,t' \leq t \leq t''\}.$$

10.3 LONGITUDINAL SEMIGROUP FOR EVOLUTION

To treat the group structure connected with time evolution, the notation introduced above has to be made more detailed, with explicit marking of the initial and final instants. Denote a path by $[q]_{t'}^{t''}$ instead of $[q]$ and a measurement result by $\alpha_{t'}^{t''}$ instead of α. All the preceding formulas can easily be rewritten in the new notation. Specifically, the evolution law will be

$$|\psi''(\alpha)\rangle_{t''} = U\left(\alpha_{t'}^{t''}\right)|\psi'\rangle_{t'}.$$

If one more continuous measurement is performed on the time interval $[t'', t''']$ giving the output $\alpha_{t''}'^{t'''}$, then at the instant t''' the system will be in the state

$$|\psi'''(\alpha'\alpha)\rangle_{t'''} = U\left(\alpha_{t''}'^{t'''}\right) U\left(\alpha_{t'}^{t''}\right)|\psi'\rangle_{t'}.$$

This state should not depend on whether we present the measurement on the time interval $[t', t''']$ as a consequence of two measurements or as a single more lengthy measurement. This means that the following equality should hold:

$$U\left(\alpha_{t''}'^{t'''}\right) U\left(\alpha_{t'}^{t''}\right) = U\left((\alpha'\alpha)_{t'}^{t'''}\right). \qquad (10.13)$$

Thus alternative measurement outputs corresponding to all possible time intervals should form a *semigroupoid*. This means that the product of two alternatives is defined provided they correspond to neighbouring time intervals:

$$\alpha_{t''}'^{t'''} \alpha_{t'}^{t''} = (\alpha'\alpha)_{t'}^{t'''}.$$

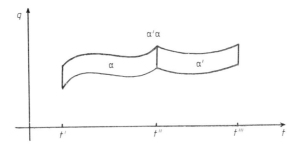

Figure 10.2: The product of measurement outputs (for position monitoring as the measurement).

For example, for position monitoring, when $\alpha = [a]$ is a corridor (around the path $[a]$), multiplication

$$[a']_{t''}^{t'''} [a]_{t'}^{t''} = ([a'][a])_{t'}^{t'''}$$

is defined (provided that $a(t'') = a'(t'')$) by the formulas (figure 10.2)

$$[a]'[a] = [a''], \qquad a''(t) = \begin{cases} a(t) & \text{for } t' \le t \le t'' \\ a'(t) & \text{for } t'' \le t \le t''' \end{cases} \qquad (10.14)$$

The same formula is valid for the multiplication of paths $[q']$, $[q]$.

The evolution operators $U\left(\alpha_{t'}^{t''}\right)$ corresponding to the measurement outputs $\alpha_{t'}^{t''}$ form, as a consequence of equation (10.13), a representation of this semigroupoid.

The prefix 'semi' here indicates that an inverse element does not generally exist in the algebraic structure under consideration. A semigroupoid differs from a semigroup in that an arbitrary pair of elements cannot be multiplied. In this case only those alternatives which correspond to adjacent time intervals can be multiplied. Some additional condition may be required for one to be able to multiply α' and α. For example, for position monitoring $[a']$ and $[a]$ can be multiplied if the end of $[a]$ coincides with the beginning of $[a']$.

Remark 1 If the Hamiltonian does not depend on time explicitly, then time is homogeneous and one may move from a semigroupoid to a semigroup. Then any elements of our group structure can be multiplied though not all of them have inverse elements. This construction will be considered in section 10.4.

The measure $d[q]$ on the space of paths and the displacement operator $V(\alpha)$ are multiplicative. Therefore the requirement (10.13) will be valid if the measurement functionals are also multiplicative:

$$w\left(\alpha'^{t'''}_{t''}, [q']^{t'''}_{t''}\right) w\left(\alpha^{t''}_{t'}, [q]^{t''}_{t'}\right) = w\left((\alpha'\alpha)^{t'''}_{t'}, ([q'][q])^{t'''}_{t'}\right).$$

The functional (10.12) satisfies this requirement if the function $W(x)$ is exponential and the functional w is Gaussian:

$$w\left([a]^{t''}_{t'}, [q]^{t''}_{t'}\right) = \exp\left(-\frac{1}{\sigma}\int_{t'}^{t''} (q-a)^2 \, dt\right). \tag{10.15}$$

In this case monitoring the coordinate of the system (or the set of all coordinates) described by the functional (10.12) possesses the property of multiplicativity. Therefore a change of the state with time can be looked at in this case as evolution of the system with the measurement taken into account. The measurement itself may be understood then as a consequence of partial measurements performed during consecutive time intervals. (See, however, Remark 2 below).

The functionals (10.10) and (10.11) are not multiplicative, and the corresponding measurements may be called integral in time. No natural longitudinal group structure exists for measurements of this type.

It can be shown that the multiplicativity property exists if the functional describing the measurement has the form

$$w\left(\alpha^{t''}_{t'}, [q]^{t''}_{t'}\right) = \exp\left(-\int_{t'}^{t''} f(A(q(t)) - a(t)) \, dt\right). \tag{10.16}$$

For $f(x) = x^2/\sigma$ one has a Gaussian functional describing continuous monitoring of the observable $A(q)$. In the case of an arbitrary (positive) function f the functional (10.16) also describes monitoring of $A(q)$ but with the help of another (non-Gaussian) measuring device. Thus monitoring is in fact the most general case of evolutionary (local in time) continuous measurement.

Remark 2 Actually the Gaussian characteristic (10.15) of the device is not appropriate for describing the time evolution of a continuously measured system because it leads to an effective measurement error

$$\Delta a = \frac{\sigma}{t'' - t'}$$

depending on the time interval $T = t'' - t'$. There is no problem if only one time interval $(t'' - t')$ is considered. If the time intervals are different (as for the description of time evolution) one has a choice: (1) take σ constant, providing multiplicativity of the weight functional but with an effective error depending on time, or (2) take σ proportional to the time interval, providing constant effective error but sacrificing multiplicativity. Instead of this one can choose the functional (10.16) with the function $f(x)$ in the form of a 'potential well' with a flat bottom. The limiting (ideal) case is a function of the form

$$f(x) = \begin{cases} 0 & \text{for } |x| \leq \Delta a \\ \infty & \text{for } |x| > \Delta a \end{cases}$$

corresponding to a rectangular characteristic of the measuring device. This choice is better than Gaussian since it provides both multiplicativity and the constant measurement error. We have used Gaussian monitoring in the preceding chapters because of its mathematical simplicity. This seems correct for estimations up to the order of magnitude for the given time interval.

So far we have described continuous measurements with the help of integrals over paths $[q]$ in configuration space. In general this is insufficient and the integral over paths $[q, p]$ in the phase space is needed. The evolution operator then has the form

$$U(\alpha_{t'}^{t''}) = \int d\left([q]_{t'}^{t''}\right) d\left([p]_{t'}^{t''}\right)$$

$$\times \exp\left(\frac{i}{\hbar} \int_{t'}^{t''} (p\dot{q} - H(p,q))\, dt\right) w\left(\alpha_{t'}^{t''}, [q,p]_{t'}^{t''}\right) V[q]. \quad (10.17)$$

Multiplicative functionals describing evolutionary measurements have the form (for $\alpha = [a]$)

$$w\left(\alpha_{t'}^{t''}, [q,p]_{t'}^{t''}\right) = \exp\left(-\int_{t'}^{t''} f\big(A(q(t), p(t)) - a(t)\big)\, dt\right). \quad (10.18)$$

Note that the introduction of such a functional is formally equivalent to the inclusion of an imaginary term in a Hamiltonian.

10.4 UNIFICATION OF TRANSVERSE AND LONGITUDINAL STRUCTURES

Mensky (1983b, 1985b) has constructed an extension of the semigroup of trajectories (parametrized paths) in analogy with the extension of the translation group to the Galilei group. It was shown that such an extension can be used for a group-theoretical derivation of a path integral and the action for a nonrelativistic particle in an external field. An analogous construction can be used to unify the above defined transverse group and longitudinal semigroup into a single semigroup of Galilean type. We shall do this for coordinate monitoring as an example of continuous measurement.[1]

First of all let us make use of the homogeneity of time and move from a semigroupoid to a semigroup. For $\alpha_{t'}^{t''} = [a]_{t'}^{t''}$ let us introduce the function $[u]_T = [\dot{a}]_{t'}^{t''}$ where $T = t'' - t'$ and

$$u(t) = \dot{a}(t + t'), \qquad 0 \le t \le T.$$

Now let us identify (that is treat as equivalent) the following elements

$$[a]_{t'}^{t''} \sim [a']_{t'+\tau}^{t''+\tau}$$

if they correspond to the same function $[u]_T$, i.e. if

$$\dot{a}(t) = \dot{a}'(t + \tau).$$

The result is that any two equivalence classes can be multiplied, and the semigroupoid changes into a semigroup. Indeed, for multiplication of any two classes it is enough to choose a representative in each of them in such a way that multiplication of these representatives in the sense of equation (10.14) is possible. The product gives a representative for the new class (defined as a product of classes). Elements of the resulting semigroup can be represented by curves $[u]_T$.

In addition to this semigroup, let us also introduce the group consisting of the curves

$$[v] = \{v(t) \mid -\infty \le t \le \infty\}. \qquad (10.19)$$

The elements corresponding to such curves will be analogies of Galilei transformations but with velocity depending on time. Let the configuration space possess some symmetry described by the group R (analogous to the rotation group for the configuration space of a nonrelativistic particle). Accept then

$$r[v]r^{-1} = [rv], \qquad r[u]_T r^{-1} = [ru]_T \qquad (10.20)$$

for any $r \in R$. Accepting one more relation,

$$[v][u]_T = [u + \tilde{v}]_T[v], \qquad \tilde{v}(t) = v(t - T), \qquad (10.21)$$

one converts the set G of all elements of the form $g = [u]_T r[v]$ into a semigroup.

The subgroup of elements of the form $g = [v]$ can evidently be considered as a transverse group acting on alternative measurement results (or equivalently on trajectories $[u]_T$):

$$[v] : [u]_T \to [u + v]_T.$$

---

[1] Actually nothing important is changed if an arbitrary observable A is monitored instead of the coordinate.

This action can be expressed directly in terms of paths $[a]_{t'}^{t''}$:

$$[v] : \quad [a]_{t'}^{t''} \to [a']_{t'}^{t''}, \qquad a'(t) = a(t) + \int_0^{t-t'} v(s)\, ds.$$

Therefore the semigroup G contains both the transverse group as a subgroup and the longitudinal semigroup as a subsemigroup. On the other hand, the structure of the semigroup G is quite analogous to the structure of the semigroup introduced by Mensky (1983b, 1985b) with the aim of a group-theoretical derivation of the path-integral formalism and dynamics of a nonrelativistic particle.

10.5 CONCLUDING REMARKS

An evident problem arising from the above consideration is to find which representations of the group-theoretical constructions introduced above arise in describing the evolution of the corresponding systems. A quantum system under continuous measurement exhibits evolution of a quite new type, having classical and quantum features simultaneously. This is why the representation theory analysis of a system subject to monitoring of some observable seems to be very interesting, as does a comparison of it with the corresponding analysis for a system in an external field. The Galilean type semigroups are in fact very similar in these two cases, but the representations arising in describing the evolution must be more complicated in the presence of a measurement.

11

Paths and Measurements: Further Development

In the preceding chapters the application of the path-integral formalism to quantum continuous measurements has been analysed in many different aspects. Let us now look at the problem from a wider point of view and formulate some perspectives in this direction. The aim of this concluding chapter is to show the peculiar role played by paths in quantum physics.

We shall see that, besides being an efficient mathematical formalism, paths form a powerful conceptual tool of quantum theory. As a result some paradoxical features of quantum mechanics become clearer or more consistent when formulated in terms of paths.

Three questions will be considered in this connection: (1) the relation between quantum and classical, (2) the paradox of the two-slit experiment and (3) wave–particle dualism.

It turns out that restricted path integrals provide a description of quantum processes which possesses both quantum and classical features and allows a natural transition to classical and quantum limits (section 11.1). The two-slit experiment loses its paradoxical features if it is analysed in terms of 'extended points', i.e. points together with paths leading to them (section 11.2). Finally, the description of elementary particles in terms of paths allows one to take into account equally naturally both wave (global) and corpuscular (local) properties (section 11.3).

The latter point can be developed significantly in the framework of the relativistic theory on the basis of the group of paths. This provides an elegant and powerful group-theoretical basis for the most important areas of modern quantum field theory: gravity and gauge theory.

In concluding section 11.4 the phenomenon of wavefunction collapse (or state reduction) accompanying measurement will be discussed in the framework of the path-integral approach. The complex conceptual problems of quantum measurement theory will be mentioned in this connection.

168

11.1 QUANTUM AND CLASSICAL FEATURES IN CONTIN-
UOUS MEASUREMENT

In analysing different situations we saw that the time development of a quantum system with continuous measurement taken into account can be described by the measurement amplitude, or conditional (restricted) propagator:

$$U_\alpha(q'', q') = \int_{q'}^{q''} \mathrm{d}[q] \exp\left(\frac{\mathrm{i}}{\hbar} S[q]\right) w_\alpha[q]. \qquad (11.1)$$

Here α is the output of the measurement and w_α is a functional describing the restriction of path integration corresponding to information contained in this output.

For the analysis one may understand α as the output of the coordinate monitoring (measurement of path) of a particle or of some other system. Then α can be identified with some path $[a]$ (emerging as a result of measurement) and the functional $w_\alpha[q]$ is nonzero for paths $[q]$ close to $[a]$. Integration in equation (11.1) is then effectively limited by some corridor of paths around $[a]$. The width of the corridor Δa is nothing but an error of the coordinate monitoring.

We saw in section 4.3 of Chapter 4 that the evolution of the measured system can be described by the formula

$$|\psi'_\alpha\rangle = U_\alpha|\psi\rangle. \qquad (11.2)$$

Let us now analyse these formulas from the point of view of the relation between quantum and classical features of the system.

The measurement output α is essentially classical because it arises as a result of the action of a classical measuring device. For example, in the case of coordinate monitoring the output $[a]$ is a trajectory of a particle (or a trajectory of a quantum system in its configuration space). In fact we have not just a single path $[a]$ but a corridor of paths around $[a]$. However, this is not because a real particle is quantum but because real measurement has a finite precision (and therefore finite error Δa).

The restricted (conditional) propagator U_α is a typical quantum object, an operator acting on the wavefunction of a particle (or of some other system). Thus we have here both quantum and classical features of motion, and the description includes both quantum and classical elements. The result of the evolution of the system may be described as its classical trajectory $[a]$. However, we know also that the wavefunction of the system, being originally $|\psi\rangle$, becomes $|\psi'_\alpha\rangle$ after the measurement.

It was shown in Chapters 4 and 9 that in typical cases there are two radically different regimes of measurement, classical and quantum. The classical regime arises if the measurement is rough enough ($\Delta\alpha \gg \Delta\alpha_{\text{opt}}$) and the quantum regime is characteristic of fine measurement ($\Delta\alpha \ll \Delta\alpha_{\text{opt}}$). Between these two regimes is the optimal regime of measurement ($\Delta\alpha \simeq \Delta\alpha_{\text{opt}}$) giving maximum information. Consider these regimes from the point of view of the relation between the quantum and classical elements of the system's motion.

The measurement output α cannot be predicted unambiguously. There is always some variance in these outputs, $\delta\alpha$. In the classical regime of measurement ($\Delta\alpha \gg \Delta\alpha_{\text{opt}}$) this variance coincides with the error of measurement,

$$\delta\alpha = \Delta\alpha.$$

For example, the trajectory of a particle [a] obtained as a result of measurement in the classical regime coincides with the classical trajectory [q_{class}] within the limits of the measurement error Δa. Thus the properties of the measurement results α are essentially classical in the classical regime of measurement.

What can be said about the evolution of the wavefunction in the classical regime? It was emphasized in section 9.2 of Chapter 9 that the restricted propagator U_α coincides in the classical regime of measurement with the unrestricted Feynman propagator U, i.e. with the ordinary quantum-mechanical evolution operator. Thus in the classical regime the measurement result (for example the trajectory of a particle) is of a purely classical character, while the system wavefunction obeys a purely quantum evolution law (the Schrödinger equation).

In the quantum regime of measurement ($\Delta\alpha \ll \Delta\alpha_{\text{opt}}$) the results of the measurement α display a much wider variance than the error of measurement,

$$\delta\alpha = \frac{\Delta\alpha_{\text{opt}}^2}{\Delta\alpha} \gg \Delta\alpha.$$

Thus purely classical characteristic of the motion α (for example the trajectory of a particle) is strongly influenced by the quantum properties of the system. This may be called quantum measurement noise. In the case of position monitoring the measured trajectory [a] may deviate from the classical trajectory [q_{class}] by much more than the measurement error,

$$\delta a = \frac{\Delta a_{\text{opt}}^2}{\Delta a} \gg \Delta a.$$

In the limit $\Delta a \to 0$ all paths [a] have the same probability.

What can be said about the quantum element of description, the propagator U_α, in the quantum regime of measurement? The most interesting point is that instead of a single unitary evolution operator U (as in the classical regime) we now have the set of partial evolution operators, U_α, where α can be any measurement output from the set of probable outputs. Instead of the conventional unitarity condition, the partial evolution operators must satisfy the generalized unitarity condition of the form

$$\sum_\alpha U_\alpha^\dagger U_\alpha = 1.$$

Nothing concrete can be said about each of the operators U_α in the general case of the quantum regime of measurement. Generally their behaviour is rather complicated (see for example section 4.3 of Chapter 4 and section 5.4 of Chapter 5). It is easy, however, to consider the limit $\Delta a \to 0$ of the coordinate monitoring. In this case a very narrow corridor around some path $[a]$ is obtained as a result of the measurement. One may say that in practice one path is found as a result of the measurement instead of a set of paths. According to this, just one path $[a]$ (more precisely, a very narrow corridor around it) contributes to the path integral (11.1). The only action of the resulting propagator on the wavefunction (except for changing its norm) is then to change its phase by the value $S[a]/\hbar$,

$$|\psi'_{[a]}\rangle \sim e^{\frac{i}{\hbar} S[a]} |\psi\rangle.$$

11.2 THE TWO-SLIT EXPERIMENT

The well-known two-slit experiment has already been discussed (in section 1.2 of Chapter 1) as one of the starting points of the whole path-integral approach to quantum measurements. This experiment is a classic illustration of the peculiarities of quantum mechanics. There are some paradoxical features in this experiment. They are that the picture P_{12} appearing on the scintillation screen when both slits are open is not a simple sum of the pictures P_1 and P_2 seen when the first or second slit is open.

This contradicts the classical view that the particle can reach some point x of the screen only through one of the slits (1 or 2), and thus the corresponding probabilities p_1, p_2 must be added to give the probability p_{12} of reaching the screen with both slits open.

Of course this paradox arises only because the situation has been considered classically. We know the quantum procedure leading to the correct results, which consists in applying amplitudes instead of probabilities. It was discussed in detail in section 1.2 of Chapter 1.

However, we shall try to avoid the paradox in another way. Look for the origin of the paradox. It is that we have *just one point* x but different

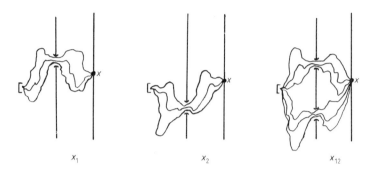

Figure 11.1: Extended points for interpretation of the two-slit experiment.

paths for reaching it. Evidently there would be no reason to try to add the probabilities if we were to consider different points in the situations denoted as 1, 2 and 12. The paradox could not arise in this case.

Let us therefore try a new approach. Let us talk not about points on the screen but about *extended points*. Each extended point consists of the point itself and all possible paths leading to it. Then in three versions of the two-slit experiment we have one ordinary point x but three *different extended points* x_1, x_2 and x_{12}. The extended point x_1 consists of the point x with all paths leading to this point through the first slit. The extended point x_2 consists of the same point x but with the paths that lead to it through the second slit. Finally the extended point x_{12} consists of the point x with the paths leading to it through both slits. (figure 11.1). Then there is no reason why the probabilities corresponding to the first two extended points should be added to give that corresponding to the third.

This solution of the paradox seems at first glance to be unnatural because its essence is almost tautological: it boils down to a different way of saying the same thing. However, this renaming corresponds in our opinion to the spirit of quantum mechanics. The main idea in this renaming is that the *target of a quantum process cannot be considered separately from the process itself*. The target (the point on the screen) has to be considered together with the process leading to it (the paths leading to this point). Only this combination (summarized in the image of an extended point) forms an *adequate language* for analysing the phenomenon in question.

The same idea can be formulated in another way: it is more natural in quantum mechanics to deal with *processes instead of states*. Of course, normal practice is to consider states. However, there seems to be a tendency to move to the language of processes instead of the language of states. One argument in support of this is that any state is formed as a result of some process or other.

It seems that the concept of an extended point accepted above is not a transition to processes instead of states but to considering processes (paths) *together* with states (points). However, in accordance with the preceding paragraph, the target point x can of course be represented adequately by paths leading to it. Therefore we may understand the extended points x_1, x_2, x_{12} as the corresponding *sets of paths* (leading to the point x through respectively slit 1, slit 2 or both slits). The reader may see already that this final image is close to the sets of paths α introduced in this book for analysing continuous measurements.

The advantage of the language of paths as a tool for solving the two-slit paradox is more evident if one takes into consideration a possibility such as the delayed choice considered by Miller and Wheeler (1984). The point is that the two-slit experiment can be modified in a special way. In this modification two different arrangements of the experiment are possible. In one the slit used by the particle is known after the experiment is over. In the other it is impossible to know this. It is only much later than the instant at which a particle enters a slit that the decision is made about whether the measurement should be performed according to one scheme or the other. Quantum interference effects arise only in the case when the slit used is not known. This corresponds to what we discussed in section 1.2 of Chapter 1.

A delayed choice allows one to choose between interfering and noninterfering schemes at the very last moment. It is evident in this case that the whole history of a particle on its way to a target point is important for analysis of the phenomenon. Thus only an extended point, the target together with the set of histories, is suitable for analysis.

The general conclusion that can be drawn from this argument is that the *language of paths* is preferable to the language of points in quantum theory. There are two different ways of using paths in quantum theory. One, discussed in detail in this book, is the Feynman path integral. The other is the formalism of path-dependent wavefunctions (or fields) proposed by Mandelstam (1962). Both ways are rich and beautiful mathematically and physically, but only the second has convinced the author that paths are better than points for quantum theory. We shall touch on this slightly in section 11.3.

11.3 WAVE–PARTICLE DUALISM

The situation with two slits is one illustration of wave–particle dualism typical of quantum mechanics. The essence of this dualism is that there are two aspects of quantum objects, in which they resemble a wave or a particle in the classical sense. Neither of these aspects can be abandoned; there are situations in which each of them becomes important.

The analysis of the two-slit experiment in section 11.2 suggests that the language of sets of paths could clarify some features of the wave–particle dualism. Here we shall consider this question in more detail.

Let us take for our analysis the most typical continuous measurement, the monitoring of position of (say) a particle. As has been argued in section 11.1, the result of such a measurement is represented by the trajectory $[a]$. Thus the particle obtains the typical corpuscular characteristic of position at a definite instant, $a(t)$. The precision of the measurement, or its error Δa, determines the precision with which this characteristic is known. This type of continuous measurement therefore gives us a picture of a particle, not a wave.

However, we see as a result of the analysis (see section 11.1) that the measurement gives results not characteristic of classical theory. The measured trajectory $[a]$ may differ from the classical trajectory $[q_{\text{class}}]$ of a particle by values much greater than the measurement error Δa. Thus our object, though described by the same mathematical tools as a particle (position, trajectory), is not a classical particle. We can judge about this when observing its motion.

Moreover, to describe this object completely, its wavefunction, or the state vector, $|\psi\rangle$, is necessary. This presents the wave aspect of the object. When observing the particle-like behaviour of the object (its measured trajectory $[a]$), we obtain results that depend on its wave characteristics before the observation (its initial state $|\psi\rangle$).

But the most complete picture of the interrelation between the corpuscular and wave properties of a quantum particle is expressed by the restricted (conditional, or partial) propagator $U_{[a]}$. In fact the discussion in section 11.1 of the interrelation between the quantum and classical properties of a particle is nothing but a representation (in different terms) of its wave–particle dualism.

The partial propagator $U_{[a]}$ (and therefore the evolution of a quantum particle) has both quantum and classical (wave and corpuscular) features. As a propagator (evolution operator), it is quantum and describes the propagation of a wave. However, this quantum evolution is possible only in the framework of a definite classical alternative $[a]$. Each of these alternatives is described by classical (corpuscular) terms (a trajectory) and their statistics is purely classical (probabilities but not amplitudes).

One may say that the path-integral theory of continuous measurements presents a corpuscular picture of a particle's motion along a trajectory, but in each actual evolution, for a given trajectory, quantum evolution acts: the wave propagates in a tube (corridor) corresponding to the trajectory.

We see thus that the wave–particle dualism of quantum objects may be adequately represented with the help of the formalism of restricted path integrals. The whole of this book illustrates this fact. However, there is another application of paths in quantum theory that also has something to do with wave–particle dualism. We mean the formalism of wavefunctions depending on paths instead of points.

The arguments of section 11.2 show that path-dependent functions may be of some help even in nonrelativistic quantum mechanics. However, it is much more evident that they are necessary in relativistic quantum field theory. Such functions were introduced by Mandelstam (1962) in the framework of quantum electrodynamics and generalized to the case of non-Abelian gauge fields by Białynicki-Birula (1963). The goal was the construction of an explicitly gauge-invariant formalism.

The path-dependent formalism turned out to be mathematically natural and physically adequate. One result of this is that a group-theoretical version of this formalism is possible (see Mensky 1983a). We cannot discuss this here in more detail, but we should remark that there is a deep connection between Feynman's path integrals and Mandelstam's path-dependent functions. A preliminary analysis of this connection has been made by Mensky (1983a, b, 1985b, 1990c),[1] but further development is needed. One may expect surprises in this investigation, shedding new light on the nature of quantum theory.

11.4 CONCEPTUAL PROBLEMS

The starting point for the discussion in this book is the Feynman path integral. Restriction of the region of integration has been shown to describe the influence of the measuring device on the quantum system subject to continuous measurement. Such an influence is usually referred to as reduction of the system state or collapse of its wavefunction. In our case we mean continuous reduction corresponding to continuous measurement. However, the general conceptual problems that usually arise in connection with state reduction are shared by this type of measurement too. Let us discuss them briefly (or rather mention just some of them).

11.4.1 *Continuous State Reduction*

In conventional quantum measurement theory (see Chapter 2) state reduction is taken into account by the special procedure of von Neumann's projection or some generalization of it. Instead of this, in the present approach the continuous state reduction resulting from continuous measurement is taken into account automatically by restriction of the path

---

[1] Some aspects of this formalism in connection with quantum continuous measurements have been discussed in Chapter 10.

integral. This is one of the most important advantages of the path-integral approach: the same procedure of integral restriction gives both the probability distribution of the measurement outputs and the reduction of the state under the influence of this measurement.

Of course one should keep in mind that this type of continuous reduction (like any other) is an idealization. The influence of a real measurement device may differ from this idealized influence. The restriction of the path integral suggests implicitly that the influence of the measurement is in some sense the minimal influence necessary to obtain the given type of information. The same information can be obtained from a poorer device that has a stronger influence on the system. However, it is impossible to obtain the same information with a smaller influence.

This is why the path-integral approach results in absolute quantum restrictions on measurability. No device providing the required information (specified by the functional w_α) can have less influence than is described by restriction of the path integral. There is a class of measuring devices with just this minimum influence on the system supplying the required information (the theory says nothing about how these devices may be constructed). All other devices work less efficiently, providing less measurability than is found from the path-integral approach.

In fact, even the latter non-ideal case may be treated in the framework of the path-integral method. Indeed, if the device is non-ideal (for the given type of measurement), it is because it gives additional information about the system under investigation (see footnote 3 on page 179). Therefore this device, non-ideal in the sense of the given type of measurement, is ideal for measurement of a different type, giving more information. Such a measuring device and its influence may also be described by restriction of the path integral, but with restriction to a narrower sets of paths.

Of course, the choice of the system of functionals w_α restricting the path integral for the given system undergoing the given measurement is in practice a complicated task. The particular types of restricted path integrals analysed in this book were chosen in many cases because of their mathematical simplicity.

11.4.2 What is Collapse?

The remark made in section 11.4.1 is a technical one. It claims how state reduction (wavefunction collapse) is taken into account in the present approach. There is, however, an important and not quite clear conceptual question: what is collapse and how does it occur?

The problem is that a description of collapse (like a description of any other aspect of measurement) necessarily includes classical elements. If one thinks (as many people do) that only a quantum-mechanical description of any system is completely correct while the classical description is approximate, then the emergence of classical elements in the formalism of quantum measurement should have a purely quantum-mechanical explanation.

For this aim one may (and in fact must) include the measuring device in the consideration explicitly. This device should be described as an ordinary quantum system interacting in some way or other with the measured system. One may hope to obtain, in the framework of an ordinary quantum-mechanical analysis of these systems, the same results as in the quantum-measurement analysis including collapse of the wavefunction.

This approach, however, faces major difficulties because different states of the measuring device arise in quantum superposition instead of a mixture. The difference between these two situations is similar to the difference between quantum (interfering) and classical (incompatible) alternatives (see section 1.2 of Chapter 1). The phenomenon of decoherentization (see section 2.1 of Chapter 2) consisting in the loss of information about phases between alternative states cannot be described in the framework of unitary quantum-mechanical evolution and requires something like state reduction.

One can formulate this difficulty as follows: if the measurement device is considered as a quantum system, then the question arises of how the state of this system can be identified. In other words, the measuring device should also be measured, and instead of solving the problem we only move it elsewhere.

The real problem in this context is the investigation of decoherentization as a specific process, connecting essentially quantum (unobservable in principle) and essentially classical (observable with negligibly small back influence) systems or states. Decoherentization may be introduced into the description of the quantum system under investigation or into the description of the measuring device but in both cases this stage is typical of the quantum theory of measurements but is not part of conventional quantum mechanics.[2]

There have been many attempts to avoid this peculiar stage of decoherentization. One of the most exotic is the Everett–Wheeler many-worlds interpretation of quantum mechanics (see Everett 1957, Wheeler 1957, DeWitt and Graham 1973) in which the suggestion is accepted that different alternatives (different outputs of measurement) are all realized but in different worlds, and only one of these worlds is available for our observation. This explanation, though very interesting, does not seem quite satisfactory because instead of actually solving the problem it just converts one prob-

---

[2] The paradox of the type of Schrödinger's cat arises otherwise (however, see for example Song *et al* (1990), in which ways are proposed of generating superpositions of classically distinguishable states).

lem into another. In any case, the existence of different worlds poses at least as many questions as it solves.

There is a more convincing direction in which to search for a solution. Some authors have tried to derive the picture of decoherentization as an approximate description of a quantum-measuring device (see for example Peres 1980b, Machida and Namiki 1980, 1984, Walls *et al* 1985, Joos and Zeh 1985, Zimányi and Vladár 1986, Fukuda 1987, Maki 1988, Zurek 1981, 1982, Unruh and Zurek 1989, Namiki and Pascazio 1991). The approximation is usually justified by a characteristic feature of the device, namely that it has a vast number of degrees of freedom and possesses instability of a definite type.

Most experts now accept the picture of Zurek (1981, 1982). According to this the interaction with the measuring device results in correlation between the states of the system and the states of the device while the interaction with the environment supplies a sort of dissipation converting the superposition of these correlated states into a mixture of them (the decoherentization process).

The approach considered in this book is purely phenomenological. It presents effective tools for describing collapse arising in the course of complicated (continuous and continual) measurements, but it does not describe the mechanism of the collapse. Moreover, no explicit model of the measuring device or environment is needed for the present approach. The advantage of this is a universal (model-independent) character of the conclusions drawn from the present approach.

One possible misunderstanding should be avoided in this connection. Some people think that no problem of collapse exists at all because real measurement can be described completely and correctly in the framework of ordinary quantum mechanics (without elements characteristic of the quantum theory of measurement, such as collapse) if one takes into account both the measured and the measuring systems.

A typical situation occurs with the quantum Zeno effect and its experimental verification (see section 2.2 of Chapter 2). After publication of the paper by Itano *et al* (1990) with a description of the experimental proof of this effect a number of papers appeared (see for example Petrosky *et al* 1990) where the authors tried to show that Itano's experiment did not actually prove the quantum Zeno effect. The argument is that the results of the experiment can be predicted in the framework of conventional quantum mechanics, without quantum measurement theory, provided level 2 is considered explicitly.

This type of argument contains, in our opinion, a misunderstanding of the goal of the quantum theory of measurement. This theory describes the results of measurements performed in a quantum system and the behaviour of the system under the influence of these measurements. This description is achieved in the framework of quantum measurement theory *without in-*

troducing an explicit model of the measuring device. However, one can of course consider a model of the measuring device explicitly. If this is done correctly, the predictions will be the same. In the above-mentioned case level 2 is nothing but an explicit model of the measuring device. Of course, taking this level into account allows one to obtain correct predictions for the given experiment. The quantum measurement theory does the same without explicit consideration of level 2. There is no contradiction between these two approaches.[3]

One may think then that there is no need for a quantum theory of measurement. In a sense, there is indeed no need. Any particular system in any regime of measurement can be considered in the framework of conventional quantum mechanics without using the methods of the quantum theory of measurement. However, the quantum theory of measurement allows one to consider some situations in great generality and reach some general conclusions which hardly can be achieved by analysing particular situations. Moreover, some problems of a general character cannot even be formulated without the tools of the quantum theory of measurement. Consideration of the motion of photons in Heisenberg's γ-microscope leads to the same conclusions as the uncertainty principle, but the latter is very important because of its generality as an effective instrument of physical analysis.

> The final conclusion from this analysis is that the problem of collapse can be avoided in practical calculations but inevitably emerges on the theoretical (conceptual) level. However, even in practical investigations the quantum theory of measurement with its picture of wavefunction collapse is useful (in fact necessary) firstly for simplifying computations and secondly for obtaining general results of the type of the uncertainty principle. Obtaining general conclusions of this type is one of the goals of modern quantum measurement theory (see for example Martens and de Muynck 1990).[4]

---

[3] It is important, however, to note the remark of Petrosky *et al* (1990) that Itano's experiment does not completely correspond to the ideal situation of the Zeno effect, since it leads to an additional correlation between states of the atom and photons radiated.

[4] The hypothesis is sometimes put forward that the consciousness of an observer plays an important role in the phenomenon of state reduction (collapse). (This idea began with the founders of quantum mechanics. Of modern papers let us mention an interesting one by Squires (1988).) The motivation for investigations in this direction is the apparent impossibility of solving the conceptual problem of collapse and the resulting conclusion that quantum mechanics is not closed in itself. We cannot discuss this hypothesis here, though it is useful to think about arguments on this topic.

11.4.3 *Quantum–Classical Evolution*

In one way or another, measurement leads to the emergence of classical elements in the description of a quantum system. In the path-integral approach this is seen in modification of the evolution law (see section 11.1 above and section 4.3 of Chapter 4). With the measurement taken into account evolution is described by the set of (alternative) operators U_α where α is the measurement output. The purely classical element α arises in this description as a sign of the influence of the measuring device. This may be treated in a wider manner as the influence of some sort of measuring medium (see section 4.4 of Chapter 4).

In fact a new type of evolution law is introduced in the framework of the present approach. The question then arises (connected with the discussion of section 11.4.2) of what is the status of this quantum–classical evolution law.

It is evident that some sort of measurement is always performed in a real situation. Indeed, measurement in our sense is nothing other than obtaining information about the state of the system and putting it in classical form so that it can be read without further disturbance of the system. However, such information is always present (for example a rough estimation of the position of an elementary particle). One may say that any quantum system moves in a measuring medium of some sort or other.

On the other hand, the influence of this medium in some particular situation may be negligible so that the ordinary quantum-mechanical evolution law is valid. This situation was analysed in section 11.1.

It is important that the boundary between the quantum system and the measuring device (medium) can be introduced in different places. Let us imagine for simplicity that there is a chain of systems $S_0, S_1, S_2, \ldots$, each measuring the preceding one. Then one may consider just one system S_0 explicitly, taking all the others into account by restricting the path integral of the system S_0. The influence of the measurement (expressed by the restriction of the path integral) will then be comparatively large. Instead of this one can treat the (interacting) systems S_0, S_1 explicitly, and take all the others into account by restricting the path integral of this compound system. Then the influence of measurement may be less. Finally one may consider explicitly sufficiently many subsystems so that the influence of all the others is negligible. Then the purely quantum-mechanical description becomes correct.[5] This is a specific formulation of a general argument in

---

[5] However, in this case one cannot know in principle which alternative is realized from the set of quantum alternatives describing the system. We have a purely quantum but unobservable system.

section 11.4.2.

It may be said therefore that the description of a real system in the framework of the path-integral theory of continuous measurements is an approximation to reality. The convenience of this approximation depends on the aim of the investigation. The arguments of the preceding chapters show that in some cases this approximation may lead to useful conclusions about some general features of quantum systems.

An opposite and intriguing point of view is also possible (see for example Ghirardi *et al* 1988), namely that continuous reduction is a necessary counterpart in the description of quantum systems. This means that collapse is supposed to occur spontaneously, without any special device or medium to cause it. The theory of quantum systems in this case should differ from conventional quantum mechanics, including some features of quantum measurement theory as its internal features. For example, one may think that the restriction of path integrals is a necessary part of the right theory.

Let us note in conclusion that the restriction of path integrals allows one to naturally overcome purely mathematical difficulties in their definitions (see Mensky 1992d). Indeed, the restriction of a nonrelativistic path integral to a corridor (even one arbitrarily wide) in coordinates and velocities makes this integral a mathematically well defined object, unlike the original Feynman path integral. If a restriction in velocity turns out to be insufficient, a restriction in higher time derivatives may be added. It is evident that a restriction of this type can be readily justified as describing a real physical situation when it is known in advance that the coordinates and their derivatives cannot be arbitrarily large.

References

Accardi L 1981 *Phys. Rep.* **77** 169

Accardi L 1984 Some trends and problems in quantum probability *Quantum Probability and Applications to the Quantum Theory of Irreversible Processes* ed L Accardi *et al* (*Springer Lecture Notes in Mathematics 1055*) (Berlin: Springer) p 1

Adachi S, Toda M and Ikeda K 1989 *J. Phys. A: Math. Gen.* **22** 3291

Anderson A and Anderson S B 1990 *Ann. Phys., NY* **199** 155

Barchielli A, Lanz L and Prosperi G M 1982 *Nuovo Cimento* B **72** 79

Bergmann P G and Smith G J 1982 *Gen. Rel. Grav.* **14** 1131

Białynicki-Birula I 1963 *Bull. Acad. Polon. Sci. Sér. Sci. Math. Astron. Phys.* **11** 135

Bin Kang Cheng 1989 *Phys. Lett.* **135A** 70

Bloch I and Burba D A 1974 *Phys. Rev.* D **10** 3206

Blokhintsev D 1982 *Space and Time in Microworld* (Moscow: Nauka) (in Russian)

Blokhintsev D 1987 *Principal Questions of Quantum Mechanics* (Moscow: Nauka) (in Russian)

Bohm D 1952 *Quantum Theory* (Englewood Cliffs, NJ: Prentice-Hall)

Bohr N and Rosenfeld L 1933 *K. Danske Vidensk. Selsk. Mat.-Fys. Meddr.* **12** N8 3–65

Braginsky V B 1967 *Zh. Eksp. Teor. Fiz.* **53** 1434

Braginsky V B 1977 *Topics in Theoretical and Experimental Gravitation Physics* ed V DeSabbata and J Weber (New York: Plenum) p 105

Braginsky V B and Vorontsov Yu I 1974 *Usp. Fiz. Nauk* **114** 41 (1975 *Sov. Phys. Usp.* **17** 644)

Braginsky V B, Vorontsov Yu I and Khalili F Ya 1977 *Zh. Eksp. Teor. Fiz.* **73** 1340

Braginsky V B, Vorontsov Yu I and Khalili F Ya 1978 *Pis'ma Zh. Eksp. Teor. Fiz.* **27** 296

Braginsky V B, Vorontsov Yu I and Thorne K S 1980 *Science* **209** 547

Busch P, Casinelli G and Lahti P J 1990 *Found. Phys.* **20** 757

Busch P and Lahti P J 1990 *Ann. Phys., Lpz.* **47** 369

Calzetta E 1989 *Phys. Rev.* D **40** 380

Casati G, Chirikov B V, Izrailev F M and Ford J 1979 *Stochastic Behavior in Classical and Quantum Hamiltonian Systems* ed G Casati and J Ford (New York: Springer) p 334

Castagnino M A 1988 The appearance of time in quantum gravity *Quantum Gravity (Proc. 4th Seminar on Quantum Gravity, Moscow, 25–29 May 1987)* ed M A Markov, V A Berezin and V P Frolov (Singapore: World Scientific) p 269

Castrigiano D P L and Mutze U 1984 *Phys. Rev.* A **30** 2210

Caves C M *et al* 1980 *Rev. Mod. Phys.* **52** 341

Caves C M 1983 *Quantum Optics, Experimental Gravity and Measurement Theory* ed P Meystre and M O Scully (New York: Plenum) p 567

Caves C M 1985 *Phys. Rev. Lett.* **54** 2465

Caves C M 1986 *Phys. Rev.* D **33** 1643

Caves C M 1987 *Phys. Rev.* D **35** 1815

Chirikov B V, Izrailev F M and Shepelyansky D L 1981 *Sov. Sci. Rev.* C **2** 209

Chiu C B, Sudarshan E C G and Misra B 1977 *Phys. Rev.* D **16** 520

Cook R J 1988 *Phys. Scr.* **T21** 49

Davies E B 1976 *Quantum Theory of Open Systems* (London: Academic)

d'Espagnat B 1976 *Conceptual Foundations of Quantum Mechanics* 2nd edn (Reading, MA: Addison-Wesley, Benjamin)

d'Espagnat B 1983 *In Search of Reality* (New York: Springer)

DeWitt B S 1964 Dynamical theory of groups and fields *Relativity Groups and Topology* (London: Blackie) p 598

DeWitt B S 1967 *Phys. Rev.* **160** 1113

DeWitt B S and Graham N 1973 *The Many-Worlds Interpretation of Quantum Mechanics* (Princeton, NJ: Princeton University Press)

Dirac P A M 1958 *Principles of Quantum Mechanics* (Oxford: Clarendon)

Dirac P A M 1972 *Fields and Quanta* **3** 139

Duru I H 1984 *Phys. Rev.* D **30** 2121

Duru I H and Kleinert H 1982 *Fortschr. Phys.* **30** 401

Everett H III 1957 *Rev. Mod. Phys.* **29** 454

Feynman R P 1948 *Rev. Mod. Phys.* **20** 367

Feynman R P and Hibbs A R 1965 *Quantum Mechanics and Path Integrals* (New York: McGraw-Hill)

Fukuda R 1987 *Phys. Rev.* A **35** 8; A **36** 3023

Gell-Mann M and Hartle J B 1990 *Complexity, Entropy, and the Physics of Information, SFI Studies in the Sciences of Complexity* vol VIII, ed W Zurek (Reading, MA: Addison Wesley) or see *Proc. 3rd Int. Symp. on the Foundations of Quantum Mechanics in the Light of New Technology* ed S Kobayashi *et al* (Tokyo: Physical Society of Japan)

Ghirardi G C, Omero C, Weber T and Rimini A 1979 *Nuovo Cimento* A **52** 421

Ghirardi G C, Rimini A and Weber T 1988 *Found. Phys.* **18** 1

Golubtsova G A and Mensky M B 1989 *Int. J. Mod. Phys.* A **4** 2733

Grishchuk L P 1989 *Quantum Effects in Cosmology, Lectures presented to VI Brazilian School on Cosmology and Gravitation, Rio de Janeiro, July 1989*

Grosche C 1989 *J. Phys. A: Math. Gen.* **22** 5073

Halliwell J J 1988 *Phys. Rev.* D **38** 2468

Halliwell J J 1989 *Phys. Rev.* D **39** 2912

Halliwell J J and Hawking S W 1985 *Phys. Rev.* D **31** 1777

Hartle J B 1988 *Phys. Rev.* D **38** 2985

Hartle J B and Hawking S W 1983 *Phys. Rev.* D **28** 2960

Helstrom C W 1976 *Quantum Detection and Estimation Theory* (New York: Academic)

Holevo A S 1982 *Probabilistic and Statistical Aspects of Quantum Theory* (Amsterdam: North-Holland)

Home D and Whitaker M A B 1986 *J. Phys. A: Math. Gen.* **19** 1847

Itano W M, Heinzen D J, Bollinger J J and Wineland D J 1990 *Phys. Rev.* A **41** 2295

Itzykson C and Zuber J-B 1980 *Quantum Field Theory* (New York: McGraw-Hill)

Joos E 1984 *Phys. Rev.* D **29** 1626

Joos E 1986 *Phys. Lett.* **116A** 6

Joos E and Zeh H D 1985 *Z. Phys.* B **59** 223

Kac M 1958 *Probability and Related Topics in Physical Sciences* (New York: Interscience)

Kaempffer F A 1965 *Concepts in Quantum Mechanics* (New York: Academic)

Khalili F Ya 1981 *Vestnik Mosk. Universiteta, Ser. 3, Phys. Astron.* **22** 37

Kiefer C 1987 *Class. Quantum Grav.* **4** 1369

Klauder J R 1987 *Recent Developments in Mathematical Physics (Proc. 26th Int. Universitätswochen Kernphys., Schladming, 17–27 February 1987)* Berlin, p 80

Kraus K 1981 *Found. Phys.* **11** 547

Kraus K 1983 *States, Effects and Operations* (Berlin: Springer)

Landau L D and Lifshits E M 1958 *Quantum Mechanics, Nonrelativistic Theory* (Oxford: Pergamon)

Landau L and Peierls R 1931 *Z. Phys.* **69** 65

Larsen U 1986 *Phys. Lett.* **114A** 359

Machida S and Namiki M 1980 *Prog. Theor. Phys.* **63** 1457, 1833

Machida S and Namiki M 1984 *Proc. Int. Symp. on Foundations of Quantum Mechanics* ed S Kamefuchi et al (Tokyo: Physical Society of Japan) pp 127, 136

Maki Z 1988 *Prog. Theor. Phys.* **79** 313

Mandelstam S 1962 *Ann. Phys., NY* **19** 1

Martens H and de Muynck W 1990 *Found. Phys.* **20** 257 355

Mensky M B 1976 *Induced Representations Method: Space-Time and Concept of Particles* (Moscow: Nauka) (in Russian)

Mensky M B 1979a *Phys. Rev.* D **20** 384

Mensky M B 1979b *Zh. Eksp. Teor. Fiz.* **77** 1326 (Engl. transl. 1979 *Sov. Phys.-JETP* **50** 667)

Mensky M B 1983a *The Path Group: Measurements, Fields, Particles* (Moscow: Nauka) (in Russian; Japanese translation Kyoto: Yosioka, 1988)

Mensky M B 1983b *Teor. Mat. Fiz.* **57** 217

Mensky M B 1985a On quantum theory of measurements of gravitational field *Proc. 3rd Seminar on Quantum Gravity, Moscow, 23–25 October 1984* ed M A Markov, V A Berezin and V P Frolov (Singapore: World Scientific) p 188

Mensky M B 1985b A group-theoretical approach to a path integral *Group Theoretical Methods in Physics (Proc. 2nd Int. Seminar, Zvenigorod, 24–26 November 1982)* vol 1, ed M A Markov, V I Man'ko and A E Shabad (London: Harwood) pp 613–30

Mensky M B 1988a *Theor. Math. Phys.* **75** 357

Mensky M B 1988b *Ann. Phys., Lpz.* **45** 215

Mensky M B 1989 *Theor. Math. Phys.* **80** 689

Mensky M B 1990a *Class. Quantum Grav.* **7** 2317

Mensky M B 1990b *Phys. Lett.* **150A** 331

Mensky M B 1990c Application of the path group in gauge theory, gravitation and string theory *Gauge Theories of Fundamental Interactions (Proc. XXXII Semester in Banach Int. Math. Center, Warsaw, 1988)* (Singapore: World Scientific) p 395

Mensky M B 1991 *Phys. Lett.* **155A** 229

Mensky M B 1992a *Phys. Lett.* **162A** 219

Mensky M B 1992b *Found. Phys.* **22** 1173

Mensky M B 1992c *Phys. Lett.* **169A** 403

Mensky M B 1992d *Teor. Mat. Fiz.* **93** 264

Mensky M B, Onofrio R and Presilla C 1991 *Phys. Lett.* **161A** 236

Miller W A and Wheeler J A 1984 *Proc. Int. Symp. on Foundations of Quantum Mechanics* ed S Kamefuchi *et al* (Tokyo: Physical Society of Japan) p 140

Misra B and Sudarshan E C G 1977 *J. Math. Phys.* **18** 756

Morette-DeWitt C 1972 *Commun. Math. Phys* **28** 47

Namiki M and Pascazio S 1991 *Phys. Rev.* A **44** 39

Newton T and Wigner E P 1949 *Rev. Mod. Phys.* **21** 400

Padmanabhan T 1989 *Phys. Rev.* D **39** 2924

Peres A 1980a *Am. J. Phys.* **48** 931

Peres A 1980b *Phys. Rev.* D **22** 879

Petrosky T, Tasaki S and Prigogine I 1990 *Phys. Lett.* **151A** 109

Piron C 1976 *Foundations of Quantum Mechanics* (Reading, MA: Benjamin)

Presilla C 1990 *Doctoral Thesis* University of Rome (unpublished)

Presilla C, Jona-Lasinio G and Capasso F 1991 *Phys. Rev.* B **43** 5200

Press W H, Flannery B P, Teukolsky S A and Vetterling W T 1986 *Numerical Recipes: The Art of Scientific Computing* (Cambridge: Cambridge University Press)

Rietdijk C W 1987 *Found. Phys.* **17** 297

Rosenfeld L 1966 *Entstehung, Entwicklung und Perspectiven der Einsteinschen Gravitationstheorie* ed H-J Treder (Berlin: Akademie-Verlag)

Shepelyansky D L 1983 *Physica* D **8** 208

Song S, Caves C M and Yurke B 1990 *Phys. Rev.* A **41** 5261

Squires E J 1988 *Found. Phys. Lett.* **1** 13

Thorne K S, Drever R W P, Caves C M, Zimmermann M and Sandberg V D 1978 *Phys. Rev. Lett.* **40** 667

Unruh W G 1979 *Phys. Rev.* D **19** 2888

Unruh W G 1989 *Phys. Rev.* D **40** 1048

Unruh W G and Zurek W H 1989 *Phys. Rev.* D **40** 1071

von Borzeszkowski H-H 1982 *Found. Phys.* **12** 633

von Borzeszkowski H-H and Treder H-J 1988 *The Meaning of Quantum Gravity* (Dordrecht: Reidel)

von Neumann J V 1932 *Mathematische Grundlagen der Quantenmechanik* (Berlin: Springer) (Engl. transl.: *Mathematical Foundations of Quantum Mechanics* (Princeton, NJ: Princeton University Press, 1955)

Vorontsov Yu I 1981 *Usp. Fiz. Nauk* **133** 351

Walls D F, Collet M J and Millburn G J 1985 *Phys. Rev.* D **32** 3208

Wheeler J A 1957 *Rev. Mod. Phys.* **29** 463

Wheeler J A 1968 Superspace and the nature of quantum geometrodynamics *Battelle Rencontres* ed C DeWitt and J A Wheeler (Reading, MA: Benjamin)

Wightman A S 1962 *Rev. Mod. Phys.* **34** 845

Wigner E P 1939 *Ann. Math.* **40** 149

Wigner E P 1968 *Am. J. Phys.* **38** 1005

Zeh H D 1986 *Phys. Lett.* **116A** 9

Zeh H D 1988a *Found. Phys.* **18** 723

Zeh H D 1988b *Phys. Lett.* **126A** 311

Zeh H D 1989 *The Physical Basis of the Direction of Time* (Berlin: Springer)

Zimányi G T and Vladár K 1986 *Phys. Rev.* A **34** 3496

Zurek W H 1981 *Phys. Rev.* D **24** 1516

Zurek W H 1982 *Phys. Rev.* D **26** 1862

Index